More things people are saying (or just saying in general) about The Physics of Video Games:

"In high school I was told analyzing video games was a waste of time and brain power. Dan proves them wrong. You hear that [almost any teacher] YOU'RE EXISTENCE AS A SCIENCE TEACHING IS A LIE!" - Mike Anthony *Still Loading*

"This is a book" - Scott Ludwig

"So... when are we gonna start that Pro Street Fighting League, dude?" - Jeremy Pierce *Gaming Futurist*

"I like trucks" - Leo 'Laserfrog' Wichtowski *Run! Play! Think!*

"Let's burn it!" - Eric Laughton

"Sweet! Awesome read, I approve of the rocket fuel explanation, I would not have thought of that. We're more than happy to have a section devoted to Cloudberry Kingdom, in fact that's freaking awesome." - Jordan Fisher *Pwnee Studios* (in reference to the Infinite Lives section)

Website: www.facebook.com/PhysicsofVideoGames
E-mail: physicsofvideogames@hotmail.com

INTRODUCTION

Welcome to the second edition of *The Physics of Video Games*. This is an expansion of the original book, with extra content squeezed in to make it a more cohesive and diverse package than before. There are articles which were featured on RetrowareTV, as well as new sections only to be seen in this book and never anywhere else. Many of these additions are based on suggestions presented to me, thus I give credit to those individuals both in the respective section and the credits, please check out their work where applicable.

For those who read the first book, you can see that a lot has remained in tact and the general structure is the same. So to be friendly to those who are reading this again, in the credits, anything that is italicized is a new addition and is where you should guide your attention to. For those who did not read the first book, I have refined the original formula (editing and eradication of pictures to reduce costs) and added a lot more content for your enjoyment. A lot of the new content follows less rigorous calculations unlike the original material. This allows for more freedom in hypothesizing solutions, coming up with explanations and presenting quick mentions of games so as many suggestions as possible can be integrated throughout the text.

The problems which we examine touch base on many games throughout the decades. I am aware not every person has played every game, or some even any. So I want to give enough insight into the game to provide context, while spoiling none or very little story elements as there are amazing twists and situations in many of these games.

I hope you enjoy the book as we get to shout people off buildings, climb colossi, potentially electrocute a squirrel-like creature, take down a helicopter, teleport, yank on spines, make chocolate float, soar majestically and many more fun activities. Just keep in mind while reading, nopody is berfect. I am very open to criticisms, enjoy! Thank you very much for your support, this has been an absolutely wonderful journey that I hope to continue! Let us try and make the seemingly impossible possible together with these hypothetical situations.

THE PHYSICS OF VIDEO GAMES PART II
THE LOST LEVELS

TABLE OF CONTENTS

SECTION I – CLASSICAL MECHANICS
FORCES AND MOTION – FUS RO DAH!

Video games contain many genres which divide audiences of gamers seeing as not everyone will love everything. The first genre to be analyzed is the platforming game which has garnered less attention in more recent years, but was the most popular type of game in the 90s. The idea of a platforming game is quite simple, the player will take control of a character or characters and jump around in environments to reach some sort of an end goal. Some platformers will contain enemies which must be vanquished along the way, others choose total avoidance. No matter what route the developers choose, the player will need to reach their end goal safely by the means provided.

Nintendo's Super Mario Bros., Konami's Contra, Capcom's Mega man and many other platforming games are very guilty of having characters which have the ability to change the direction of their jump in midair. This is certainly not a complaint as it is one of the many reasons these games are as incredibly difficult as they are. Without this mechanic, they would be infuriating. Seeing as the whole idea behind a platforming game is to be able to traverse environments while avoiding enemies and hazards; only being able to go forward would promote the player to frequently missing their platform and falling into the bed of spikes below instead.

A part of the reason people are capable of jumping is because gravity is pulling them towards the surface of the Earth. Without gravity and atmosphere people would be floating around and not be able to propel themselves in any given direction. This leads to an important point, the difference between weight and mass. Mass is the total composition of an object. Let us say a person named Batyodi is in a universe, he is going to be composed of the same amount of mass no matter where he is (unless he forgot to repay a favour to the mafia). Batyodi's weight however, will be constantly changing as he moves through the universe. If on Earth, his weight could be 100kg, but on a different planet he could weigh 30kg due to a lower acceleration of gravity.

The next fact we need to consider is Newton's third law which speaks about every action having an equal and opposite reaction. When people compress their legs and pounce, they are pushing down on the Earth, since the Earth is a pretty stable surface, people are able to propel themselves due to the force of the Earth pushing back up on them. When a person does jump it is gravity that pulls them back down again. Depending on the way the person jumps, they travel straight up and down, or in a forward arching motion; many platforming games do not agree.

As was stated before, many games allow the player to change his or her's in game character's direction in midair. We will examine Contra (as it is probably the worst offender of the three previously mentioned games) which has two soldiers who are pitted against hordes of aliens across various terrains. Their jump heights are approximately three times the height of the respective character. This is not an unreasonable occurrence, which may come as a surprise to some people. The Contra series takes place on Earth which has the average acceleration of gravity $9.8\text{m}/s^2$, but the soldiers are experienced at battling aliens. This means there is a chance these soldiers have been fighting these aliens on different planets which have very strong forces due to gravity. Thus, the soldier's muscles have grown stronger and adapted to the different planet's conditions. Upon returning to the Earth, gravity would comparatively have less of a restraining effect on the soldiers much like when an astronaut goes to the moon and jumps on its surface. The moon's gravity is much lower than the Earth's, so an Earth dweller on the moon would be able to jump much higher. This is assuming the soldiers had devices which allowed for safe travel on planets with much stronger forces of gravity as the human body has certain limits. Without the aids the soldiers would look like pancakes much like cartoon characters constantly switching directions in an elevator.

As a brief segue we will examine what the value of gravity actually dictates. We have already established the average acceleration of gravity on Earth is $9.8 \text{m}/s^2$, but we need to examine the units closely. We have metres per second squared, which means that every second that goes by, whoever is falling towards falling towards the Earth will speed up by an extra 9.8m/s. So if someone is initially starting a free fall, after the first second they would move at 9.8m/s, the next second at 19.6m/s, the next at 29.4m/s and so forth (neglecting air resistance). It is always important to think about what the units given actually mean. Back to our brave soldiers.

No matter what planet these soldiers or any of the previously mentioned characters are on, changing the direction in the midst of a jump is physically impossible. They do not have the freedom to go where they want, when they want at any point during a jump. As was stated, there needs to be some type of surface which will provide enough force to produce an opposing force to allow the person to jump. For this example, we will consider the soldier from Contra, Bill Rizer, to be on the Earth. When our dear Bill Rizer jumps forward, then suddenly changes direction to return to the vertical distance in which he started, there is a major problem. The air in the Earth's atmosphere does not provide a substantial enough opposing force to allow Bill to propel in the opposite direction (so this completely discredits double jumping as well). There is resistance in the air which slows forward motion, but not enough to allow him to rapidly change direction. The only way this could ever occur is if there were extremely focused air packets which forced Bill to change direction in midair. There is a beloved Capcom (we may have mentioned in the beginning) character that might be able to overcome this hurdle though.

Maybe Mega Man with all the technological advances, could change directions in midair; maybe in the year 200X, there is a technology which allows Mega Man to overcome this problem. Mega Man looks harmless enough, but I am very certain that if any person had him running toward them shooting lemon lasers and jumping ridiculous heights, he or she would be terrified. Mega Man has amazing abilities which allow him to overcome countless obstacles. One of these technological abilities may just be the key to explaining how the Blue Bomber can change directions in the midst of a jump. In the robotic, complex design of his blue suit, hidden far down in unseen pixels are potentially high powered little vacuums. These vacuums create air packets that act like platforms that can be used similar to a springboards, allowing for midair change in travel. If Mega Man's suit was made from a material not yet discovered here in our world there is another possibility.

A material so light and so durable, that changing directions needs very little force. Gundams are made of

Gundanium (I hate you at times anime). For arguments sake let us assume Mega Man is made of Megamanarium (I guess I am not any better). Megamanarium will be a very complex solid composed of fundamental particles rather than protons, neutrons and electrons, thus meaning it will have extremely small mass and will be held together by strong, electromagnetic and nuclear forces. Magamanarium is even more fundamental that leptons and quarks and is protected be a thin, but durable exoskeleton with materials found on the periodic table. It is hereby my super material discovered today, explains how Mega Man is actually the most complex physics phenomena to strike us ever. For he is made of particles impossible to discover by today's standards. Maybe, the year 200X is actually 20010 and one day we will be that advanced. Maybe Capcom is actually a team of the most brilliant scientists ever to grace this universe who know that fate of our world. Maybe, there is a definite answer to the meaning of life. Or maybe Capcom was just merciful for once with these painfully difficult games.

Similar in principle to changing directions, let's talk about a Batman game. Who would have thought that Sunsoft and Batman put together would have been such a wonderful combination? If we aren't keeping this question rhetorical I suppose the obvious answer would be the developers knew. Well-paced difficulty progression, short with no filler, infinite continues, perfect controls, dark and diverse atmospheres, incredible music, what is not to love? I know Batman has been in a lot of rubbish and we have people like Frank Miller and Joel Shumacher to thank for that. But trust me when I say that this game is proper tidy. But what to examine? (Read this next paragraph as if Laser-Frog of *Run, Play, Think!* were narrating. Check out his work if you are not familiar.)

We could examine how Batman punches people until they explode, how batarangs can split from one to three of equal size or how seemingly infinite amounts of supplies can fit in one belt. But those all can be answered in one sentence. That wouldn't be any fun now would it? Me just answering by saying: robots with self-destruct buttons on their chests, three spring loaded blades connected together and it's a utility belt so get over it. No. You would probably be sad the article was only two sentences. You would say "why is the explanation only two sentences? I don't want it to be two sentences. I want it to be longer. I want something to read with my non-polygonal coffee. I need more than two sentences, so I can finish my non-polygonal coffee". That wasn't any fun, just getting it over with like that... well time to end this since I am not funny like LaserFrog. Let us get right into examining the main mechanic driving the platforming sections of the game: being able to jump off walls.

Similar to the way Mega Man changes directions in midair, being able to change directions applies all the same to Batman, but rather than vacuum packets in the air, we are talking about jumping off walls. A utility belt can only do so much, so let us see how the Caped Crusader can jump off walls so readily whilst obeying the laws of physics (mainly that pesky gravity).

You may have seen videos of people who run up walls for a considerable and impressive distance until gravity ruins the vertical progression party. These individuals are using friction and momentum to their advantage in order to climb high, but gravity will always triumph no matter how much skill in in place while climbing. Yet there is the slower paced ice climber who can climb seemingly insurmountable vertical stretches using nothing but a series of spikes in strategic locations (some intense climbers do this with no safety lines!). This means that Batman needs to create some sort of a hybrid of wall-runner and ice climber in order to progress in his quest to stop the Joker.

With all of the money Batman has it is quite feasible that he could invest in and create spring-loaded spike boots. Wayne Enterprises has certainly invested in weirder and more expensive endeavours like Bat-credit cards (I try to be positive, but I seriously hate that movie). Sorry to those who did not know, I may have just spoiled that Bruce Wayne is Batman for you. At least these boots have a practical application and a wonderful name which we will dub to be: Batsproots (that name is worse than the credit card I know). What these Batsproots do is utilize tiny spikes which plant themselves into a surface at a slight angle in order to overcome gravity for a short amount of time. Since gravity is working against Batman, he needs very powerful springs in the boots along with a guiding push from his hand to launch him off of the surface he is on.

This seems to be the case as Batman is only able to remain affixed to a wall briefly prior to falling. This means he should be able to climb endlessly so long as he maintains a decent pace. Perhaps we should not have given away Batman and Sunsoft's secret, now any building which has extremely high, narrow shafts with electrically charged blocks as exits might get robbed. Now let's used everything we've learned thus far and apply it to another Capcom game: Mega Man X (X not being a roman numeral for 10, it is also the name of the character this time around).

MMX is one of those occasions where amazing eats a

lot of brilliant for breakfast. MMX was a title that expanded on the ideas presented in the original MM franchise and refined others to make an extremely enjoyable game. The intense music, diverse and inventive weapons, outstanding boss battles, everything about this game is proper tidy.

The mechanic in MMX we are going to draw our attention to is one just introduced to the MM universe in this title: wall jumping. MMX has two additions that make it different from the way Batman jumps off of walls. First is the ability to climb up a single wall and make vertical progress. Secondly there is the ability to remain affixed to a wall while sliding down said wall.

The first addition can be explained using what we explored with MM in the original series and the ability to change direction midair. X has the ability to create a series of air packets to push off of in order to return to the wall that was launched off of, but at a different height. This means that X will be able to continue to make upwards progress using a single wall, but only if we understand the second ability.

The second addition of being able to slide down the wall rather than plummeting into a free fall can be explained using the vacuums as well. X uses the vacuums as the equivalent of a series of suction cups, meaning remaining affixed to a wall as definitely possible. The sliding downwards will occur because there is enough force to allow X to remain on the surface of the wall, but not enough to become motionless due to the weight of the bulky armor. With all of our groundwork laid out with two other games, this explanation fell out rather effortlessly as an extension. Although it may not seem like it, many of us physicists are lazy (optimists may call it efficiency), so we like to take the easiest route to explain something.

Since we are having fun with air-packets, we'll apply it to a long running series created and published by Bethesda called the Elder Scrolls with its fifth installment in 2011 called, Skyrim. This installment is a very open and expansive game which very impressively showcases what is known as a sandbox RPG. An RPG (role playing game) is typically thought as a game where players assume the role of a character, or characters, in a fictitious world where they are able to train and become a formidable foe to those around. As an example, the character may have a base strength of one, and after defeating enough enemies, he or she may gain experience to have a strength level of two. This could also just be refering to the character gaining abilities as the game progresses whether it be to jump higher or shoot fireballs.

The term "sandbox" means players have the freedom to explore the world around them with little to no restrictions creating a story and character unique to each player experiencing the game. Many games follow a linear, restricted fashion which involves the player going from point A to point B with little deviation. The linear progression can create an incredible experience if done well, but ultimately every player will have an experience similar to everyone else playing. A well thought-out sandbox game, such as Skyrim, allows every player to explore their game in an order unique from every other player, thus creating discussion amongst players and a sense of ownership of the play through experienced. Now back to the packets of air.

Skyrim contains a variety of ways to approach combat with an enemy whether it be bashing a skeleton with a club, slicing at a dragon with a long sword, or shooting a fireball at a merchant innocently running a shop in his or her hometown. The player's character eventually gains an ability called "Dragon Shout" which allows the character to utilize a variety of fictitious words to enhance the character's abilities or harm an enemy. The technique we are going to focus on is a dragon shout which when cast has the playable character shout "Fus Ro Dah!" thus creating an intense air packet which launches foes a considerable distance considering.

Let us detour for a moment to get an understanding of what sound is. The definition of sound varies depending on whether one views it from a philosophical or a scientific standpoint. The former believes that sound is a sensation experienced by a person, while the latter believes it is the vibration of particles in the air which occurs whether it is heard or not. This is why the old saying "If a tree falls in the woods and there is no one there to hear it, does it make a sound?" still causes debates to this day. We will be going by the notion that sound is not merely just a sensation, but the vibration of particles in the air.

Particles in the air can come under an enormous pressure if compressed by certain sources such as a rocket taking off or a car racing across the desert in an attempt to break the sound barrier. The pressure exerted from these objects has the force to move objects around them and even perforate someone's eardrums. The particles of air will become very tightly packed into one another around the front of say a speeding car. Now for the sake of argument, we will believe in magic for the time being and that someone can shout so loud (without ruining their vocal chords) that they can actually lift someone off the ground and launch them like a projectile using their dragon shout. Let us try to determine what sort of force this wave can provide.

We can always determine the force of something from the simple equation: F = ma; where F is the force in newtons (N), m is the the mass in kilograms (kg) and a is the acceleration. So a newton would be the net force used in order to accelerate a mass of one kilogram by one m/s^2. In our previous topic with the soldiers on Earth, we could use their weight to determine the amount of force they would exert on the surface of the Earth using our knowledge of the average acceleration of the gravity being 9.8m/s^2. Now in the case of the dragon shout, we will have two unknowns to solve for. First we need to know what the acceleration of this shout is, then we need to know what its mass is. This introduces an unique challenge; how can we measure the weight of air inside of a video game? It is simple, we can't. Even if this were to occur in real life, we can not simply take the amount of air particles we have in the area of the dragon shout and put them on a scale. But, we can at least calculate what velocity he hits the ground at as well as the velocity in which he is launched at as a projectile.

Let's assume that our character is out and about and sees a guard standing near the ledge of a tall watchtower; this seems like the perfect opportunity to valiantly yell "FUS RO DAH!" to his back to push him off the ledge and watch him plummet to the ground below. This situation in-volves what is known as projectile motion, where the object (the guard in this case) moves forward while being forced down towards the ground via gravity. This is an ideal situ-ation which we picked this for a number of reasons. Firstly, if we have a longer time between when the initial shout was cast and the time the guard hits the ground, we can get a measurement which would involve a little more negligible error than if we did this on a flat surface. The reason for this is that starting and stopping our timer at the exact moment when the guard was launched and when he landed, respectively, is very difficult to do. If we give ourselves a longer period of flight, we have more of a chance to neglect those fractions of a second difference in those start and stop times because of how small they will be in comparison to the overall flight. Secondly, we would like to neglect the friction (which we will examine in several sections) of the guard sliding along the ground, as this is a force which we do not have the means to calculate.

When we honourably shout at this guard's back, we need to enforce two conditions. First, we need to ensure that our shout is horizontal (which we will call our x axis), so both the shout and guard are not working against gravity. We would like for our guard's motion to be caused only on the x axis thus reducing the number of variables we need to account for in our equation. Second, we need to ensure we are as close to the guard as possible so that we can eliminate hitting the guard at an angle with only a part of the shout. Now that we have those conditions established, we need to make a few assumptions in order to proceed with the calculation and trial.

Seeing as the whole world of Skyrim is obviously fictitious, we need to make a few assumptions. Even if the approximations we make are not correct, the process will nonetheless be the same. The game also does not claim to be on our Earth, but we will do these calculations assuming it is. As far as we know, Skyrim could take place in a world where everyone is 100 metres tall and weigh 2700kg. For the sake of relevance we will assume his height is 1.8m falling at a rate of $9.8 \text{m}/s^2$. We now have enough information to set up our trial to sacrifice a virtual life in the name of science.

Our guard friend innocently stood by his post while our 'hero' attacked with the dragon shout and let the guard plummet to his doom. From his honourable sacrifice (which he was unaware of his commitment to), we gathered the following data: he travelled horizontal and vertical distances of 9.0m and 5.4m respectively for 3.0 seconds. To gather the data, the guard's height used as a measuring device and a stop watch timed the whole ordeal. These are approximations of course, but we can gain some valuable information from these. As stated before, whether the numbers are correct or not, the process remains the same, so we could do this multiple times to yield more accurate results.

This is where us shouting strictly on the x axis is beneficial since there is only one factor we have to consider: the speed in which the guard is travelling due to the force of the dragon shout. So we can calculate this from the distance and time the guard travelled in the following way:

$$v_x = \frac{d_x}{t} = \frac{9.0\text{m}}{3.0\text{s}} = 3.0\text{m}/s\,[\,forward\,]$$

Where v is the velocity on the x axis (hence the x subscript), d is the distance and t is the time. We need to understand the difference between speed and velocity. Velocity has a direction, which is why we specified the forward direction. Speed is a magnitude, meaning that is has no direction. So our result means that every second that passes the guard will soar 3 metres through the air along the x axis, this is our velocity. If we were concerned with speed, we would say the guard is travelling at 3.0m/s. Since we are concerned with direction, we cannot consider this a speed. Saying the guard is travelling 3.0m/s could refer to him moving upwards, downwards and anywhere in between. We use velocity seeing as the direction of travel is important. This is an average velocity over the course of the whole trip. If we considered air resistance the speed would constantly be changing. So for the sake of argument, we will neglect air resistance.

We are also able to determine what velocity the guard hit his head in the downward direction using our time of flight and acceleration in the following way:

$$v_y = a_y t = (9.8\text{m}/s^2)(3.0\text{s}) = 29.4\text{m}/s\,[down]$$

Where v is the velocity in the y (vertical dimension) and a is the acceleration due to gravity. Something to take notice of is that no matter how far we let this guard soar majestically, the speed in the x dimension will never change assuming we have no air resistance seeing as there are no external forces. The velocity in the y dimension will be different at every moment in time due to our dear friend having force of gravity causing him to accelerate. We now have enough information to calculate the overall velocity which the guard hits his head at.

The values which we calculated have a specific magnitude and direction, so we can call them vectors. The vectors are at a 90 degree angle, or perpendicular, to one another; so using their magnitude we can think of them as sides to a right angle triangle. Using the pythagorean theorem we can calculate the hypotaneuse or in our case, the overall speed:

$$v^2 = v_x^2 + v_y^2 = (3.0\text{m}/s)^2 + (29.4\text{m}/s)^2 \rightarrow v = 30\text{m}/s$$

Now to make this speed a velocity using trigonometry:

$$\tan\theta = \frac{29.4}{3.0} \rightarrow \theta = 84.17^\circ \rightarrow v = 30\text{m}/s\,[84^\circ\ below\ horizontal]$$

Now is a good time to discuss significant figures.

Calculating our angle and final velocity, we had more decimal places than what is written down as the final answer. We round the final answer to the amount of significant figures we have with our initial values. We only knew two digits in our initial values which we used whether they be the times or distances. If we kept say, four figures in our final value, that would mean our final value is more accurate than our initial values. This cannot be the case, so we need to round our values to the same amount of figures we had in our initial values. Also note, we need to be as accurate as the value with the least amount of figures which we use. Even if we knew our time to seven decimal places, if our distance only had two figures, then the distance would dictate the amount of figures.

Going back to the velocity, this is quite an incredible speed to be hitting the ground at! We can analyze this result to see if our values are in the correct range. When the guard's face pressed itself upon the cobblestone on the ground below at this point in the trip, it appeared as if he was almost falling completely downward. That was what we could see in the trial, which completely matches the results. 84 degrees below the horizontal indicates there was a velocity along the x axis, but it is nothing compared to the velocity in which he was falling. Our guard friend will be missed, but his efforts were not wasted from all of the data we gathered. We will continue examining forces and motion with a short look at gelatinous, over-sized breasts.

Developed by Team Ninja and published by Tecmo Koei, Ninja Gaiden Sigma 2 (NGS2) takes the player on an unusual adventure with a strange breed of action game. The player is faced with cripplingly difficult fights with perfect controls leaving no one but the player to blame for the one thousandth death on the same enemy. There is a massive weapon inventory and ability set with anything ranging from a giant scythe, to fireballs, to an incredibly long flail. There are four different characters to play as: main character Ryu and three other women. There is a large amount of content to examine and mention from this game, but we will narrow our sights to one unusual concept: breast shaking using nothing but the player making motions with the controller.

Whether someone views this breast shaking mechanic as sexist, funny, titillating (no pun intended), necessary, or any number of things, physics has a non-judgemental answer as to how this works. The programmers of this game included two dominant motions which the game will detect: rapidly shaking the controller straight up and down and shaking it left and right. The key word is a rapid motion. It can not be a slow, gradual movement. To explain this we can return to our simple formula of $F = ma$. If we have a rapid jerking motion, there will be a great acceleration, thus a great force. A slow motion would have an incredibly slow acceleration which would not produce a great enough force for the player to jiggle any virtual breasts (this is awkward to write about, but we will persist!).

This is similar in principle to a person lifting a cup of tea. If someone lifts their cup quickly, when they stop the motion the cup and arm will stop together because they are a connected system; the tea however is not a part of this system and has a force imposed on it and a place to go. The tea is going to move in the direction of the force previously induced. The person would be safe until the force of gravity overcomes the force of the upwardly moving tea. When gravity becomes the only force, this means the person is no longer safe from hot water scolding them. If someone slowly lifts the cup, the force will be so small that when the person stops lifting the tea, there will not be a great enough force to launch the hot tea upwards and spill it all over. This is something which is probably not said often: the movement of breasts in NGS2 is the exact same principle as a person lifting a cup of tea.

If the person slowly moves the controller in either of the directions described before, there will not be a great enough force imposed on the sensor inside of the controller to register movement for the breasts. Whereas rapid movements up and down will result in bouncing breasts and rapid motion left and right will result in rigorous clapping of the breasts (seriously, this is awkward, my mother suggested I cover this topic!). It should be noted that slow movements not moving any of the female character's breasts is not due to the controller lacking sensitivity as it is quite sensitive to motions, even subtle ones. This was a design choice of Team Ninja. They may have thought that a rapid movement would create a more immersive and inviting gaming experience. The breasts on each female character are rather large which exaggerates these motions to make them more noticeable. If it is any consolation to those who may be offended by the inclusion of breast jiggling, this is not something pointed out in the game and not a primary focus by any means (probably not comforting in any respect if you're offended). This brings up another point.

A number of factors have frustrated people who cannot seem to look the other way as they are not forced to play games including: gore, drugs, alcohol, crude behaviour, strong language and the list goes on. There is one in particular which raises debates with not only games, but the media in general: the titillating features of the show-cased character. We already examined the physics involved with shaking breasts inside one particular game, but this singles out too focused an audience. We need to examine a spectrum of titillating characters made for males and fe-males alike regardless of their sexual preference. Some people may view the human body as beautiful and should be exposed, others view it as something which should be hidden. In either case, we can analyze the practicality of these situations. We will examine a comparison between different female characters, starting with someone we have already touched based with.

We will draw our attention to Rachel, one of our fe-male characters in Ninja Gaiden Sigma 2. We already know that the player has the ability to move her gigantic breasts with the controller, we will now examine whether or not her overall physique is practical. Having gigantic breasts is nothing which defies any sort of conventions. There are plenty of females with large, jiggly breasts with small waists. Our problem lies in what the rest of her tight, re-vealing outfit shows us: no muscle structure. Women with large breasts will not increase their muscle mass simply by supporting their natural (or implanted) chests pillows. This is not even the concern with no muscles. The concern lies in the fact that she fights in high heels, uses a massive ma-chine gun and a hammer almost the size of herself, all while she is doing flips, spins and attacking enemies. This is where her character design falls flat (figuratively speaking). There is no possible way for her to do all of this without having an extremely developed muscle structure. So clearly there was no thought in whether this character could exist or not. She is purely there to draw in those males, females and anyone in between, who are aroused by the equivalent of a 3D rendered broom with watermelons attached.

Developed by NetherRealm Studios and published by Warner Bros. Interactive Entertainment, Mortal Kombat is a game which pits fighters of various backgrounds against each other in the most gory, over the top way possible (going back to the controversy statement, this series is basically the birthplace of controversy in video games). Players have the ability to let a lizard man melt an opponent by spitting acid or let an officer tear a person in half while doing a handstand using nothing but their legs. This game is controversial for a number of reasons and bothers many people, but we are going to direct our attention to one character as a comparison to Rachel. There is no shortage of large breasted, under-dressed, exposed exist in Mortal Kombat (nor many should you fancy that); but we will narrow our focus to one of the large breasted women, Mileena.

Very much like Rachel, Mileena has quite a bit of attention drawn to her body, especially when it is her costume composed of stray bandages. Mileena also has to use her body quite aggressively to take on her opponents, whether it be to leap great bounds to attack them in the air, or to toss someone twice her weight well above her head. The developers put a great deal of thought put into her being sexually pleasing to some players, as well as carefully considering her practicality. The previously described activities would demand Mileena be physically fit, which she is. She has well defined and strong leg, arm and abdominal muscles which are all involved in those processes. Although both Rachel and Mileena are intended to attract certain players, Mileena has the advantage over Rachel because she has a more practical body for what she is doing. Rachel would have to be bulkier and stronger than Mileena to even begin to consider some of the things she is able to do in NGS2. At least in the defense of the clothing choice for both characters, they both need to wear something non-restrictive and being quite exposed in very little latex-like material is certainly non-restrictive.

We've focused a lot on how female characters in games are sexually exploited in realistic and unrealistic fashions. A number of games contain barely clothed men showing their bulky bodies flashing off their goods for the player, stimulating some, making some feel inadequate, others indifferent. To avoid redundancy, we will not be examining the male cases as the comparisons follow a similar fashion. There are a number of characters in which the comparisons can done with seeing as a game like Mortal Kombat, although ridiculous, has quite accurate portrayals off bulky men barely clothed, whereas many other games have barely clothed men with completely made up muscles to show their strength. The only way these fake muscles could exist are if the characters practically injected themselves with some form of plastic to form them. This obviously creates some enormous health hazards.

While keeping Mortal Kombat in mind, we will examine one of the more memorable and controversial concepts in the game, the fatality where a head can be ripped off of someone with the spine remaining in tact. Rather than describing something so gruesome and awful, we will hide this concept with some poetry (thanks to a fan under a pseudonym for the suggestion).

No1likesdennis suggested we speak
Can we rip out a spine, an answer we seek

We need two to fight
Find out what is right
Scorpion and Tsung, are they too weak?

- — -

Scorpion begins by launching his spear
Tsung quickly dodges, showing no fear
Defense is required
He counters with fire
Cast in flames, one side this battle did steer

- — -

Scorpion enraged, jumps back to his feet
He then summons flames to cook him like meat
Tsung jumps in the air
He acts from despair
Back to the ground by the foe he must beat

- — -

Blows are exchanged, up, left and right
Tsung's skull is smashed with tremendous might
Stunned and dazed he stands
Defenseless from hands
Scorpion swings hard to heighten Tsung's plight

- — -

Skull now cracked, a spear straight to the gut
Increasing the stun, Tsung deepens his rut
A large roundhouse kick
Tsung flies awfully quick
Scorpion then taunts, confident he struts

- — -

Tsung now embarassed, tried to fix his pride
Forcefully he runs with a confident stride
Exchanging their gaze
Tsung's met by a blaze

Tsung now spent, a fatality is now tried

- — -

Scorpion proclaimed "I want his spine and head!"
He grabbed at his neck, surely now he'd be dead
He pulled and he yanked
He yelled and he cranked
Tsung was now lightly bruised up instead

- — -

Not enough strength, his anger grows
Humiliation! Round two we now go
A new lease on life
They increase their strife
The potentially new victor, an honour shall bestow

- — -

Shang Tsung shouts out "your soul will be mine"
Scorpion with confidence knew he'd be fine
A loud cry he roared
He pulled out his sword
With a clean swipe, Tsung was cleft in twain

- — -

No one is able to claim another's spine
Yanking or pulling the foe will be just fine
No amount of strength
Can go that great length
But with a sharp enough blade, one can be cleft in twain

- — -

Revenge will be had by someone Tsung had created
Alleviated he'll be, rather than frustrated
Avenged he will be
Too bad he won't see
Mileena arrives completely aggrevated

- — -

One will obey physics at the very least
Mileena is quite a remorseless beast
She will not announce
When she'll take a pounce
And use tarkatan teeth to have one great feast

In case the language and limericks made the answer as clear as mud in the night, here is a summary: the initial proposition of ripping out a spine is impossible, but blades being sharp enough to cut someone in half is possible. We obviously won't test it, let's just assume it works.

FLIGHT – WE LIKE IKE WENT KABLOOIE

To address the matter of flight in games, first we will examine someone special from Nintendo's fantastic strategy RPG title released on the Wii: Fire Emblem: Radiant Dawn. We will do so by following up the limericks of Mortal Kombat with Petrarchan sonnets.

This time we'll look at Fire Emblem: Radiant Dawn
A strategy RPG, of the highest degree
The qualifier strategy is truly the key
No ability to revive an as example, once dead, they're gone
What to examine? Magic's too complex to think upon
Destruction of weapons is too difficult to see
We need to find something to think about critically
A character is what we'll seek, but not just some pawn
Ike is too obvious and but an ordinary man
Micaiah uses magic, let's ignore this great mage
The Black Knight is nice, but spoilers induce rage
We will find someone of power, a great ally
The man with the wings and becomes a hawk on demand
Tibarn is our subject, the warrior of the sky

- – -

The Hawk King Tibarn flaps his wings with power and
grace

With great confidence and strength, all others seem weak
When he sets his sights, the enemies' chances seem bleak
To excessively dominate, a new form he can embrace
When in hawk form, even more fear is on the enemies' face
A thunderous roar can let out from the limb-tearing beak
No matter the power, does he obey physics, an answer we
seek
Is it possible for Tibarn to flap his wings and hover in
place?
We need something to compare, something similar is pre-
ferred
An eagle's wings resemble Tibarn's a bit more
Yet the eagle cannot hover, but majestically soar
We need to think smaller, a bird who brings less demise
To compare we will use the delicate hummingbird
This comparison may be where the answer lies

- – -

Forces in balance, the condition in order to float
Horizontal motion, swaying is very low
The important balance is from above and below
The frequency of flap is important to note
Why the high speed causes this, we might not connote
Even so small, a slow flapping speed, they'll fly low
The balancing act is really important to know
Fast flapping will oppose gravity if the bird can devote
This usually won't happen if the bird is too large
Understanding the forces will give us our answer
The bird needs to be light, like the feet of a dancer
Although the larger might be saved by raw power
The same which allows for the fearful and strong charge

The one used by Tibarn to make all others cower

- – -

The analysis is the same, regardless of the form
Whether in hawk form with a deafening roar in the skies
Or on land, surrounded by his wingless allies
There is an answer which will hopefully inform
If Tibarn hopes to hover, he will need to conform
The raw power will not suffice, no matter how hard he tries
Gravity is too overbearing, given his size
He would need to flap faster, the hummingbird norm
Maybe he could hover, provided a continuous draft
Right under his wings to substantiate the lift
This may compensate for not being sufficiently swift
This would provide enough of a counteracting force
This might be too idealistic, we might be too daft
But it gives us an answer, so show know remorse

Now for those who are not exactly a fan of the poetry, let us look at flight with a different game in more practical language and format.

Banjo-Kazooie was developed by Rare and published by Nintendo in 1998. Banjo-Kazooie is a platforming-adventure game where the player will take control of a bear named Banjo and a bird named (take a guess) Kazooie. When the player sets out on the adventure, she/he will use these two to collect things. Lots and lots of things. To go on the collection adventure, jumping and flying will be required. This means we have two things to analyze this time around: collecting things or jumping and flying. The former does not seem to have the same substance as the latter, so let's do this!

We have covered jumping a few times already, so

we're not going to add another to the list (yet). Instead we are going to offer an extension to the sonnets as suggested, because Kazooie attaches herself to Banjo and allows him to fly and glide. We will also review some of the concepts mentioned in the Fire Emblem poem as some of the ideas may be as clear as mud in the night on account of the Petrarchan sonnets.

In order for Kazooie to fly upwards or glide, there needs to be a force under her wings which counteracts gravity. In the case of flying upwards, the downward motion in conjunction with the wide wing span, pushes air downward to allow for upward movement. Recall that every force has an equal and opposite reaction, and the wide span provides enough force beneath the wings to lift Kazooie. The respectively small body certainly helps her, as well as other birds, to fly with greater ease as their body weight does not introduce as great a hindrance. Gliding works in a similar fashion as the wide-spread wings allow for greater air resistance so falling will be hindered. This can be seen by dropping a piece of printer paper which is flat in comparison to crumpling it up and dropping it from the same height. This is all well and good to let Kazooie fly, but what happens when we attach a giant honey bear to her?

Kazooie's physical strength is not of concern. She should be strong enough to accomplish this feat due to the extensive training she did to try lifting a bear. If we want to really play it safe, we can assume that she has an exoskeleton which aids her in lifting Banjo. Although this explains how she is able to lift Banjo, it does not alleviate concerns with gliding or even flying for short amounts of time.

The fat honey bear Banjo is simply too big compared

to Kazooie to allow fluid flying to be acceptable. Although we established the strength of Kazooie would be sufficient, the weight carried when compared to the slow moving and small wings is too substantial by comparison. Kazooie flying with Banjo would be like either of us trying to take flight using umbrellas in our hands.

Gliding could occur as some air resistance is created, but not enough to glide like in the game. It would probably be similar to either of us trying to glide while holding two umbrellas. Classical physics does not offer an immediate solution to how this bird can fly a bear around, but perhaps some quantum field theory can (not quite modern physics section yet, but we'll get through this!).

It is theorized by some that gravitational forces are composed of massless, elementary particles call gravitons. There are also antiparticles which have opposite characteristics of their standard counterparts. Maybe Kazooie is an extremely talented bird which discovered both and has not shared her knowledge of these particles, she just exploits her discoveries. Rather than take total advantage of her discovery and show off to everyone, maybe she flaps her wings to pretend she is flying as a facade rather than admit her manipulation of gravity. It would be nice if she shared her expertise though, it sure would help a lot of physics researchers. Then again, this analysis could be in vain and we could simply say that a witch made it all possible with magic.

Lets entertain the idea of these gravitons existing and being something that people can control. With this ability, lets apply it to the wonderful game developed by Japan Studio and released by Sony Computer Entertainment on the under-appreciated Vita title: Gravity Rush.

Gravity Rush is a very unique game in the sense that the player is able to explore the entirety of the developed environments. This means everything. Walking in town and want to walk along the walls? Want to walk along the bottom of the city floating in the sky? Gravity Rush lets the player manipulate gravity and do so at almost any given time in order to walk along walls and bottom of the city in an effortless and fluid manner. This would be possible if the manipulation of the gravitons were acceptable as suggested with Kazooie.

Kat, the main character, has limits as to how long she is able to manipulate gravity before she has drop let it take control again to recharge. This would suggest that if this manipulation is possible, it is indeed a tiresome activity. I know it exhausts me whenever I take control of gravity.

SOUND – WHAT VACUUM?

Developed by Visceral Studios and published by Electronic Arts, Dead Space is a survival-horror game which is an apt category name. The survival element refers to characters trying to survive despite the impossible odds they are faced against. This could include but is not limited to being trapped without food or weapons or fending off ghosts, monsters and people. The options are virtually limitless and to some degree, one could argue every game could be in this category, but games with a deliberate focus on survival are placed under this label. The horror element refers to the player or in game character(s) being scared by any combination of people, monsters, objects and environments. Yet again, a game will be classified with this label if it is deliberately planned to scare, not accidentally. If a person were to get scared by an in game characters Skyrim, the horror label is not appropriate since it is an accidental scare.

Dead Space is deemed a survival-horror game due to its gory and graphic content, creepy environments, psychological torture and being alone on an infested mining vessel in space. Unfortunately the main character Isaac Clarke is stuck in the middle of it all. He has to attempt to survive against monsters, called necromorphs, which would love nothing more than to make him a main feast and unpleasant dinner conversation. He is armed with futuristic engineering tools which is convenient for Isaac's survival as his repair kit acts as his savior. He even has a force gun which would follow the same type of analysis as the dragon shout in Skyrim. Our analysis follows Isaac into space on the outside of the mining ship rather than inside with his weapons though.

There is a point in the game which forces Isaac to use his suit to stomp along the outside of the ship while avoiding space debris. He has boots which adhere him to the surface of the ship which takes care of one detail. Another concern is the depletion of oxygen (which allows us fleshy life forms to survive on Earth), however there is a limit to how much oxygen his suit stores and he has to access stations to refill his suit. The final concern is when he is listening to messages on the outside of the ship warning him about the debris he could potentially be injured or killed by. Based on what we already know about sound, we need to determine whether this is possible or not. So further examination of sound is needed.

We established that sound travels by means of vibrating particles in the air. The vibrations and collisions of particles in a sense passes on the information of the original sound. There are limits to this however, the further the sound travels, the less intense the sound gets seeing as each particle in the air can not pass on all of the 'information' (energy) to the next particle. This continues on until the sound eventually dissipates. The distance the sound will travel depends on the density of the medium. A medium in this case refers to some sort of a material, element or compound which has particles in it. Air is a gaseous medium, metal is a solid medium, both allow for sound to travel. Sound will actually travel faster and easier in a metal than air. The components of a metal are more tightly packed than a gas, so the components do not need to travel as far to reach another molecule to pass on the energy. This is why if someone wants to hear if a train is approaching, it is actually easier to make the determination by listening to the sound along the track rather than in the air. This does lead to more complications if the person forgets to move out of the way though. We now need to investigate what happens to sound in the vacuum of space.

In the vacuum of space, there are no particles and there is no atmosphere. This means there is no oxygen, no other components of air on Earth or even some sort of particles that would compose a sort of 'space air'. If there are no particles in the surrounding areas, our unfortunate friend will not be able to hear anything when on the outside of the ship as sound has no means of transportation. So whether it be an asteroid belt sending off debris, him firing his weapon or necromorphs charging at him, Isaac will not know what hit him unless he is staring directly at it.

This raises an important point though: why is Isaac able to see what hits him, but not hear it? We have already determined that sound needs a medium to travel in. Both light and sound are waves; waves need a medium with particles to travel in as they rely on the vibrations of the particles to move. Light is special though as it is not only a wave, but a particle as well. This duality of light confused scientists for years not only in the discoveries involved with coming to these conclusions, but in understanding the results as well as the implications. This is why light is so special and has its own special name for its travelling particles: the photon. Isaac needs luck in any place he can find it considering his difficult circumstances, so at least the photons allow him to see what is going to potentially cause his demise while travelling along the ship's exterior.

PROJECTILE MOTION – I'M GOING TO MAKE IT

Infinity Ward and Activision developed and published Call of Duty: Modern Warfare 2. This is a fast paced first person shooter game which pits the player against impossible odds that in real life the person would have either been killed or stayed home because of how insane these mission debriefs are. There is one particular section which is incredibly interesting and is a fantastic introduction to projectile motion: using a snowmobile to jump a gorge.

Our character has the ever so troublesome task of driving away from pursuers at ridiculous speeds having to maneuver around trees, rocks and cliffs and at times even having to drive with one hand to shoot followers. The likelihood of someone being capable of doing this is slim to none. Considering the speed (which we will calculate after) is insane, being able to swerve so effortlessly with one hand without coming to a grinding halt as the snowmobile does several side flips and sends our hero rolling under it, is not likely. There are soldiers who do this sort of thing in their careers and hats off to them, but writing in the comfort of a room is risky enough for some (paper cuts hurt!). This is not our primary focus though, jumping a gorge is.

At one point our hero is driving down a hill with no pursuers and virtually no obstacles, neglecting the odd tree. At the bottom of this incline is where the gorge our hero courageously jumps is located. We are able to gather a few pieces of information from this situation: the speed which he is driving, how wide the gorge is and the vertical height at the peak of his jump. First things first, let us calculate the speed he is travelling prior to hitting the jump.

It seems this game was just asking to be analyzed as it ever so conveniently tells us the distance in metres until we reach our destination. So we will start our timer at zero at a distance of 1298m from our destination then travel 12 seconds to 436m away. Since our previous calculation of velocity we had both our initial distance and time at a reference point of zero, we need to take a slightly different approach to the calculation. We need to use an average velocity formula which goes as follows:

$$v_{av} = \frac{(d_2 - d_1)}{(t_2 - t_1)} = \frac{(-436m - (-1298m))}{(12s - 0s)} = 71.833 m/s \quad \text{or}$$

$$v_{av} = \frac{(d_2 - d_1)}{(t_2 - t_1)} = \frac{(436m - 1298m)}{(12s - 0s)} = -71.833 m/s$$

The validity of either number is very important to note. The first calculation we have said that zero (our destination) is the reference point and anything prior to this is a negative distance away. This way we come out with a positive value to use, which is handier as positives are always easier to work with. The second situation dictates the opposite saying that anything prior to our goal is a positive distance away, so our distance decreases. This gives us a negative velocity, which is perfectly legitimate and we can use it. Seeing as positives are nicer, we will use the concept that anything prior to our destination is a negative distance away. We could also not use a velocity, but a speed and both answers would be the same as we are concerned with magnitude (just the value). Unlike the previous calculation, we do not write the direction explicitly as the sign dictates the direction of travel based on our conditions.

Since vehicles (Canadian at least) use kilometres per hour and not metres per second, we will convert this value to shed some light on how fast our soldier friend is going. We know that there are 60 seconds in a minute, 60 minutes in an hour and 1000 metres in a kilometre (the prefix kilo refers to a 1000 times the base measurement). We use that information in the following way:

$$\frac{[(71.833\text{m}/s)(60\text{s}/min)(60\text{min}/h)]}{(1000\text{m}/km)}=256\text{km}/h$$

This is an alarming speed, but still a physically possible one as some specially designed snowmobiles have gone much faster than that. Seeing as he is travelling this speed on the incline, this is actually a plausible part of this scenario.

Now that we have the speed of the snowmobile, using the time it takes to jump the gorge and land again, we can calculate the distance our hero jumped. Timing our jump, the trip took approximately two and a half seconds from start to finish. With this time, neglecting air resistance, we can calculate the distance in the following way (the distance we can see in the previous diagram, with our soldier friend of exaggerated height):

$$d=vt=(71.833\text{m}/s)(2.5\text{s})=179.58\text{m}=180\text{m}$$

So now we have the distance using a calculation. Recall this game was very helpful to us since it told us how far we were travelling in metres. Using the distance from the objective before and after the death defying jump, we get the values of 436m and 257m respectively. Subtracting the values we get a distance of 179m which matches our calculated value wonderfully!

We have one final value to calculate: the vertical distance at the peak of this jump. This part is an estimate and can not be confirmed within the game itself, but is an interesting aspect of projectile moment to analyze. Seeing as a projectile's path is in the shape of a parabola, there is a peak, which would be our max vertical distance during the jump. We need to know the velocity of the snowmobile along the y-axis in order for us to find the height. We will assume the angle of the jump to be 10 degrees from the horizontal. From there we can use trigonometry to find out the velocity in the y dimension in a similar fashion to the triangle used for the calculation in Skyrim:

$$v_y = v sin\theta = (71.833 \text{m/} s) sin(10^o) = 12.47 \text{m/} s [up]$$

We did not establish which direction was positive or negative in the y dimension, so we need to include direction. With this velocity we can use a kinematic equation for projectile motion to find the height as follows:

$$y = v_y t - \frac{1}{2} g t^2 = (12.47 \text{m/} s)(1.25 \text{s}) - \frac{1}{2}(9.8 \text{m/} s^2)(1.25 \text{s})^2$$

$$7.9 [upwards]$$

The peak height appeared to be at about the half way mark of the jump, thus indicating the time used. It is not possible to determine whether this number is correct or not. But seeing as the value indicates a height which is not ludicrous like 100 metres vertical, it is in the correct range.

This is a scene which at first glance may seem quite unreasonable and not physically possible, but as we just demonstrated, there is a lot that is indeed correct about this scene. Whether a real soldier would be capable of doing this is up for debate, but physics is definitely on their side this time. Developers either did their homework on this scene to make it breathtaking and plausible at the same time, or just had an incredibly accurate fluke.

TENSION – HOOT! HOOT! FREEDOM!

Next up to plate is Treyarch and Activision's Call of Duty: Black Ops and its hatred for helicopters. There is a scene in this first person shooter involving a prisoner escaping and having the enjoyment of taking down a helicopter with a harpoon gun as his first taste of freedom. Several major issues plague this particular scene. First, what is this harpoon attached to after being fired? There are no places on the side of the helicopter for this to hook on to, so it must have punctured the side of the helicopter. If it did, then when the pilot tried to fly to safety, the harpoon would have left the way it came. For the sake of argument, we will assume that this harpoon somehow managed to attach itself to the helicopter and become a single unit.

The second concern is the attachment of the other end of the rope. We have established that this harpoon enters the helicopter and is going to stay there. This does not excuse the fact that how the other end of the rope is dealt with is never really shown. It is obvious the soldier couldn't hold onto the rope to take the helicopter down, so by magic the game allows the rope to be securely attached to a pipe. No matter how this soldier has attached this rope, whether it be glue,or a knot, there is no way that it would hold in the presence of a helicopter tugging on it. For the sake of argument, let's assume that this rope somehow manages to become one with the pipe and can not be moved, we now have a single unit of a pipe, rope and heli-copter. This still brings up the biggest concern.

What is this rope made of that it has the durability to not snap when a helicopter tugs on it? The scene clearly de-picts a rope attached to the harpoon. Whether this is a nat-ural on synthetic fibre rope is hard to determine, but since the latter has a higher breaking strength we will use this for the example. We will assume it is a polypropylene rope due to its woven structure and look (Scherrer, 2010). Based on the size of the shooter's thumb with respect to the rope, we can estimate it is about two and a half centimetres thick. If we assume a very durable fibre such as polypropylene, a rope this thick would be able to withstand approximately 5500 kg of force (Scherrer, 2010). According to the U.S. mil-itary, a helicopter this size, equipped for battle would weigh approximately 10 000 kg (U.S. Army, 2010). So the very mo-ment the rope became taut, it would instantly snap due to the force of tension, letting the helicopter fly out of the clutches of the prisoner's harpoon.

Tension is caused by a pulling force. So if there were a string that was able to let a weight of mass m hang on it, tension would be created. This can be seen by the follow-ing:

$$\Sigma F = mg + T = 0 \rightarrow T = mg$$

Where ΣF is the sum of all forces, m is the mass of the object, g is gravity and T is the force of tension. This means that the tension in the string is equal and opposite to the direction of the force of the weight. The same situation applies to the helicopter pulling on the rope, but instead of worrying about the force of gravity pulling down on a weight, we have the applied force of a helicopter. Keep in mind, the weight of the rope is certainly a force which could be applied to this equation, but compared to the weight of the helicopter and the force imposed with it, the force of the rope is negligible.

If we examine the exact moment the rope becomes taut without snapping, we can apply the same type of equation. We will also assume the the helicopter is moved slow enough that there is no other applied force other than the weight of the helicopter, so it will be similar to that of the weight on a string.

$$\Sigma F = mg + T = 0 \rightarrow T = (10\,000\text{kg})(9.8\text{m}/s^2) = 98\,000\text{N}$$

According to the weight limit of our rope, it will only be able to support 54 000N. So neglecting any extra force and just considering the weight of the helicopter, the rope could not influence the helicopter's path; it would snap instantly. The only way this would be feasible is if the rope were considerably thicker than it is, just to support the weight alone, about four and a quarter centimeters approximately. The same calculation would apply to get to this value, but this still doesn't account for the force the helicopter would exert from its movement. So unless this rope were considerably thicker and tied to the pipe and placed strategically somewhere on the helicopter ahead of time, there is no way they could have stopped this helicopter with a harpoon. This is of course if we assume that the approximations and data we obtained are indeed true. The process involved will be the same regardless of the data, but if we assumed too large a weight of the helicopter and too small a tensile strength in the rope, the result of this analysis could be entirely different. While on the subject of harpoons and tension, we have another popular franchise to discuss.

The Legend of Zelda franchise is an on-going action-adventure series which has gathered fans around for decades. They showcase sandbox games by allowing the player to assume the control of a character named Link who wanders through dungeons, towns and various lands in between in order to solve puzzles, fight enemies, meet friends and save the day. Whether it be the brilliantly orchestrated soundtracks, fantastic weapon sets, or well composed dungeon puzzles, the Zelda series is adored by fans all around the world. Although we have several titles we could examine for this section, we will be examining The Legend of Zelda: Link's Awakening and the use of one particular weapon.

The hookshot is an amazing piece of technology in terms of its capabilities. When retracted the hookshot seems to be nothing more than a metal rod with a blade at the end; when extended on the other hand, it is a useful tool which allows Link to explore areas normally restricted. The basic premise is that Link will activate the hookshot, the blade will extend along a chain and track mechanism to latch onto specified surfaces at which time the hookshot will pull Link to where the blade latched onto. This makes for a device which is vital for Link to traverse difficult terrain and go over gaps which would be impossible to jump over. We are now going to determine whether this technology could work or if it is total bunkum.

We can use our analysis of Black Ops as a direct comparison. Let us consider when Link has activated the hookshot and it has latched onto a post and is just beginning to pull him towards the post. This brings up the same ideas as the questions brought up with Black Ops. First, how is this blade attached to the post? An ice climber using picks to scale a mountain, an archer shooting an arrow into a target and a ninja throwing a shuriken in a wall, they all share similarities to Link's case. If a sharp object is given enough force, it is able to puncture surfaces. The sharper the object, the stronger the guiding force and the weaker the object being punctured, the more of a chance there is for penetration. In the case with the hookshot, the blade is launched with a certain amount of force and is able to puncture posts and treasure chests, but not something like a wall. This is reasonable as it is similar to cases in real life. Our assumptions could fall apart slightly as when the hookshot is going to pull Link towards the post it sunk itself into, it seems like it would be easier to pull itself out of the post rather than pull Link towards it. Much like with Black Ops, we will assume this becomes a single system which will not move from the puncturing point.

Our second concern as with Black Ops: what is the other end of the hookshot attached to? Unlike the mysterious wrap around of the rope in Call of Duty, we can assume Link has a sturdy grip on the hookshot by some type of holster set up to allow for ease of use. It is impossible to tell as we never see a closeup in the game and it is an overhead view. We can access the instruction manual accompanied with the game which only shows a part of the hookshot, but there is somewhat of a holster mechanism for his hand, so we can make this assumption safely. The strength required to hold on to the hookshot would be considerably large, but plausible.

The final part of this situation to compare is whether this hookshot would break due to tension the moment it started to pull Link over a gaping hole. This is actually the least of our concerns with this case seeing as it is made of metal which we can easily assume has a great deal of strength to it. Based on certain conditions, it is possible to pull down a helicopter with a rope, so it is safe to assume this hookshot has enough strength to pull a person. We can assume this to be a very light and durable alloy (composition of many metals) as the game never states what the hookshot is made of. If those were not enough areas to address, we have one last piece of the puzzle to consider before progressing.

This analysis is going to be a little bit different from the other ones thus far, as this time we are not going to be subbing in values, just creating general equations for the case at hand. This way anyone can take these equations, sub in the values they believe to be true and calculate their own results. One factor is that there is no chronology to the Legend of Zelda series (short of a few titles), so we can not use another game to get an idea of Link's height. This game is a top down view with objects not familiar to the player, so we cannot use objects as a reference height. One player could argue that this is a game that takes place on a micro scale and all characters and environments in the game could all fit onto the surface of the head of a pin. There is no argument or value which would be more correct than the other, so we will use fictitious units as reference to form general equations to fit anyone's opinion.

A similar argument applies to the case of the weights involved with this game. Even if we chose a scale for the heights that correlates to familiarity, the weights could vary vastly. If we chose the height of Link to be the average height of a person on Earth, we could assume an avarage weight as well. The same cannot be said about the weight of the hookshot and other inventory though. As was said before, the hookshot could be a variety of alloys available to create a weapon with. The hookshot could even be cast iron which is extremely heavy thus creating a more effective device to puncture a post as it would have a tremendous force involved. It could very well be made of an extremely light and durable alloy. The possibilities are endless. This is why creating general equations any person can create their own assumptions is quite powerful, there will be a sense of ownership with the final values calculated.

First we will determine the equation for the velocity of Link as he escapes death with his hookshot clearing seemingly bottomless pits. This game has a overhead view, but with a slight angle to it. With this in mind we will consider the height of Link to be considered one unit. His height from this perspective matches the height of surrounding blocks and spots all throughout the game so it is very easy to measure how many units our hookshot travels. In this game it appears that in a time of one second the hookshot can travel eight units. Due to how quickly the hookshot travels, the extension and contraction distance had to be considered to measure the time. Using that information we can now form a general equation for velocity: $v = d/t = 8\,units/s\,[\,forward\,]$.

We are also able to determine the tension and forces on the hookshot. Similar to the case with Skyrim, we are able observe the forces separately in the x and y dimensions. The y dimension would only focus on the weights of both Link and the hookshot due to the force of gravity weighing them down. We are assuming the velocity of the hookshot allows for Link so travel along the x axis faster than he can fall. The hookshot has a limited range so this is not an unreasonable assumption. This brings up an important issue which troubles many people at times.

If we set up a scenario where a person shoots a gun

completely horizontal the same time someone drops a bullet, which will hit the ground first? They will hit the ground at the same time. Even considering air resistance they should still hit the ground at the same time. This is due to gravity being the only force drawing them both towards the ground which would be constant on both regardless of their horizontal speed. Try it at home with a gun which shoots foam balls. Shoot a ball horizontally and drop a separate ball simultaneously.

We can now consider the forces along the x axis which would be tension imposed by Link and the acceleration of the hookshot from rest to the velocity of 8 units/s. Unlike the case with the helicopter in Black Ops, we cannot make the same assumptions to arrive at the same formula used in that case. So for this case we have the following equation:

$$\Sigma F_x = m_L g + m_L a_L = m_L(g + a_L)$$

Where m_L is the mass of Link (including inventory), g is the acceleration of gravity and a_L is the acceleration of Link. Now that those general equations are established, any person can make her or his own assumptions to solve for the overall force in the equation. The acceleration can by determined by dividing the velocity equation by the time it takes from Link to go from a standstill to the velocity of 8 units/s. This is a rather quick time and could be an eighth of a second, or even half a second. As stated before, depending on the preference of who is using the equation, the values for mass, gravity and distances can be assumed by the person performing the calculations. That is the power of creating a general equation, it allows for discrepancies between people creating an ownership of the final calculated value.

Eight units per second regardless of the distance measurement is incredibly quick. So it is safe to assume the hookshoot is able to pull Link over obstacles impossible to jump over. Whether this technology is feasible or not is not the debate. We will let the engineers determine whether they can create a hookshoot like this or not. Whoever attempts this at least we know physics will be backing up its use and practicality. Physics does not factor in the sanity of whoever is going to clear a gorge with a piece of metal however.

HARMONIC MOTION – WE'VE OBTAINED SPACE PIRATE

We are going to turn back the retro clock to the 1990s. Nintendo's Metroid series is revolutionary in many respects, but there are two major reasons these games are as incredible as they are. Firstly, they are very large adventure games which give the player an expansive inventory at their disposal to manage the situation our lead character is in. These games are somewhat reminiscent of a sandbox game, but at later times in each game. Initially, there is a somewhat linear progression and more emphasis is placed on puzzle solving and combat. It was a welcome change to have a side-scrolling, platforming game which had combat and adventure tied in as well. The second reason this series was a break through: the main character was a female and the developers never made an example of it. There is no damsel in distress, just a female character doing what she needs to do to get through the day.

When we focus on the third game in the series, Super Metroid, we see the main character Samus Aran is completely cover from top to bottom with an armoured suit to protect her from the harsh environments she faces on the space pirate planet she lands on. There is no exaggeration of female parts, no hair flowing, just a practical suit of unisex, futuristic armour. This suit of armour has incredible powers and adaptive features which aid her along the way. Her suit is equipped with an arm cannon which is capable of using many offensive abilities such as shooting an ice blast to freeze enemies or use a wave shot to pass through walls. She also has many features which enable her to traverse the surroundings such as a morph ball to maneuver tight corridors or speed boots to pass over rapidly collapsing floors. Our attention is going to be directed towards a different gadget, the grappling beam (although many of the other gadgets along with others will be mentioned in the extras section).

When Samus uses her grappling beam to latch onto conveniently compatible blocks on roofs, she resembles a physics phenomena we all have seen in one form or another: a pendulum. Whether it be a mass swinging back and forth on a string, or a person fiddling with their pencil by clenching it between his or her thumb and index finger and letting it swing due to boredom, a pendulum is a relatively common occurrence. We could even consider a giraffe's leg moving while walking to a certain degree. There are differences between the types though. We can consider the small mass on a string to be more of an ideal pendulum, whereas the other examples would be considered a physical pendulum. We will consider both cases eventually, but for the time being we will use the ideal pendulum as an introduction and apply it to the case of Samus swinging.

Let us take the hypothetical case of a small, point mass and attaching it to a weightless string, secure this to a horizontal surface and let it swing. This is the situation of our basic pendulum. It begs the question: why a weightless string and point mass? Returning to a person pinching a pencil between their fingers, if the person pinches the pencil at the end and rocks it back and forth versus at the middle and spinning it, the pencil motion will be different. Each of the two cases will offer a different resistance to the movement. If we have a larger object such as a broom instead in the same situations, it will offer even more resistance. These situations refer to what is known as inertia: the resistance something offers when introducing motion. We would like a small mass so that we can neglect inertia to make our situation less complicated. We will use inertia when dealing with the physical pendulum, but for the time being, let's make things easier on ourselves.

Let us imagine our ideal pendulum is hanging perpendicular to the horizontal (vertical), we now pull the mass back with the string remaining taut and let it go. If we neglect air resistance (a damping force), this mass will move continuously due to gravity. The back and forth is known as simple harmonic motion. It is considered simple due to us neglecting damping effects and using ideal conditions to analyze this pendulum. Harmonic motion refers to the pendulum moving back and forth in the same repetitive pattern of sorts. We can gather some interesting pieces of information from our ideal situation. We can predict the how many times the mass will pass a certain point in a given time, known as the frequency (which is measured in cycles per second, otherwise known as Hertz (Hz)). We can also measure the time it takes to complete one full cycle, the period (T) which is measured in seconds.

Let us use an example, say we have an ideal pendulum on Earth which has a string length of one metre. We can calculate the period and frequency assuming no external forces are added to the mass and we are neglecting any damping forces. When we do these calculations, one last assumption must be made. The reasons these equations work is because we are using a small angle approximation to complete the derivation. We are assuming the motion is approximately 15 degrees away from the vertical. First let us calculate the frequency (Keep in mind to the power of ½ is the square root of a number):

$$f=\frac{1}{(2\pi)}[\frac{g}{L}]^{(1/2)}=\frac{1}{(2\pi)}[\frac{9.8\text{m}/s^2}{1.0\text{m}}]^{(1/2)}=0.50\text{Hz}$$

Recall that frequency represents the number of cycles per second, so this calculation tells us that in one second, this will complete half of a cycle. From this we can calculate the period in a couple of ways:

$$T=2\pi[\frac{L}{g}]^{(1/2)}=2\pi[\frac{1\text{m}}{9.8\text{m}}/s^2]^{(1/2)}=2\text{s} \text{ or } T=\frac{1}{f}=\frac{1}{0.50\text{Hz}}=2.0\text{s}$$

From this we find our that our ideal pendulum with a length of one metre on Earth will take two seconds to complete a cycle due to its 0.50Hz frequency. We are keeping in mind that one complete cycle means going forward from the starting position and returning.

Going back to the case with Samus using her grappling beam to swing, we can relate this back to our simple pendulum. Our first connection can be made with our weightless string. If we have Samus using a grappling beam, we will assume that it is some type of light based beam seeing as it gives off shiny blue light. The light of the sun or laser pointer have never weighed anyone down before, so we are safe assuming the grappling beam is weightless. The technology behind this beam is well beyond any technology we have at this point, but we'll assume it works. Relating back to the point mass, we chose a point mass because we need an even weight distribution in a small area so we do not need to worry about the moment of inertia. This is where our assumptions will fall apart seeing as no matter how we look at this we can not assume Samus is a point mass. It would be a convenient assumption in the sense that the calculated values may very well be in the range of our actual values and the calculation is easier. We will not take that route, so this is where the idea of a physical pendulum and including the moment of inertia come into effect.

Inertia deals with the centre of gravity and how an object will resist motion. We have many situations that we could apply in the case of a pendulum: the giraffe and pencil examples, a weightless string with a solid sphere at the end, a crane, wind chimes and many more. We are going to assume there is no air resistance as with all of our cases so that Samus can take a break and let gravity gently rock her back and forth. Another assumption could be to assume the weightless grappling beam provides a force which overcomes air resistance and provides no more force then that. Since we have no idea what powers this technology holds, it is a safe assumption. We are also going to assume that Samus is crunched up almost in a sphere shape.

With the assumption that Samus is a sphere, setting up a diagram, and using some differential equations we can come out with the following equation (for insight as to how, a classical mechanics book such as Analytical Mechanics (Fowles & Cassiday, 2005) can help):

$$T = 2\pi\left[\frac{(l_g + r_s)}{(g)}\right]^{(1/2)}$$

In this case of a weightless grapple and a sphere, the equation is almost a mirror of the ideal case aside from the inclusion of a radius. In the equation l_g is the length of the grappling beam and r_s is the radius of the ball we assume Samus to be. We will assume the radius to be half of her approximate height, 0.85m, she will grapple a distance of 8.0m. Let us plug those values into our equation:

$$T = 2\pi \left[\frac{(8.0\text{m} + 0.85\text{m})}{(9.8\text{m}/s^2)} \right]^{(1/2)} = 6.0\text{s} \text{ and } f = \frac{1}{T} = \frac{1}{6.0\text{s}} = 0.17\text{Hz}$$

If we had a longer swing with the grappling beam, we would have a period of 6.0s. This is actually quite a bit of time if doing this constantly inside of a game buck luckily our amazing grappling beam allows us to speed up our swings to make like much easier and faster. The other thing to consider is that this game takes place on distant alien planets, so the gravity could be much greater, creating shorter swinging times. If we use our point mass formula instead we get 5.7s and 0.18Hz for our period and frequency respectively. If we were able to determine the moment of inertia for Samus using calculus, provided we had more information, and we knew the force of gravity, it would get even more accurate, but this is close enough for us. We will let Samus take a break from the duties she for ridiculous reasons agrees to and let her swing in peace. While on the subject of physical pendulums we can use a general formula to determine the period and frequency of another hero.

We are going to turn our retro clocks back even further to the late 1980s to when Capcom released their classic title, Bionic Commando. This was quite the unusual side-scrolling, platforming, shooting game in the sense that players traverse vertically and horizontally without jumping.

Instead of jumping the players get the wonderful bionic arm which lets them latch onto ledges and other objects to swing and climb to their objectives. The player assumes the role of Ladd Spencer who's role is to use his bionic arm and weapons to take down the bad guys known as Badds. So we now know the creativity and ingenuity was spent on the game mechanics and not the naming of organizations. Nonetheless, it makes for a very interesting and incredible gaming experience.

There is a general formula for calculating period if we just have one object rotating, not multiple in the form of (where I is the moment of inertia):

$$T = 2\pi \left[\frac{I}{(mgl)} \right]^{(1/2)}$$

If we return to the case of someone pinching a pencil at the end and wiggling it like a pendulum between their fingers, the same principle will apply to Mr. Spencer. His arm is a giant cylinder, his body is almost cylindrical and even his gun is a giant cylinder. So for the sake of the calculation, we are going to assume that Spencer is Cylinder Man for the time being.

As with cases before, we need to establish certain conditions for us to be able to complete the calculation. First, we will assume the same condition that we did with Samus that Cylinder Man is overcoming air resistance but not swinging with his own force, just gravity. We will assume he is 1.8m tall and has extended his arm 3.6m and a total weight of 120kg. If we are assuming Ladd is a cylinder, then we have to assume a radius that we can use. Assuming his waist is 0.18m and bionic arm 0.08m, we will pick a value slightly above halfway in between seeing as there is more body than arm, so the overall radius will be 0.15m for

Cylinder Man. Using our conditions as well as the moment of inertia for a cylinder being $ma^2/2$ (a is radius) we can calculate our period:

$$T = 2\pi \left[\frac{[(120\text{kg})(0.15\text{m})^2]}{[(120\text{kg})(9.8\text{m}/s^2)5.4\text{m}]} \right]^{(1/2)} = 0.33\text{s}$$

$$f = \frac{1}{0.33\text{s}} = 3.00\text{Hz}$$

In the game Cylinder Man swings using his arm providing extra force on top of the natural motion so we can not confirm this calculation exactly. In game he does swing quite fast and if he completes a whole cycle in a third of a second for that given length, it does confirm that our calculation is in the area we need. Luckily Ladd can swing as quick as he can so that he can save virtual humanity from the Badds.

FRICTION – BRICKS AND HAIR CAN SUPPORT ME

Since we have already introduced the idea of a sandbox game, let us discuss a game which takes a different approach to maneuvering around the land. Ubisoft's Assassin's Creed II takes an entirely different approach to the way a game like this should work; instead we get to control an assassin who can scale enormous churches, watchtowers and other buildings in many unique and interesting ways. This is where we could suggest to our dear friend Ezio that the guards got up there using stairs and ladders, but it is much more fun this way.

Assassin's Creed II takes place in an old Italy (Da Vinci makes an appearance even!) where stunning architecture is brought to life through the power of computer programming. Crowded streets with beggars, prophets and street vendors are abundant. Along with this, large buildings which Ezio needs to get a better vantage point, escape trailing soldiers or even find his victim are in no shortage either. Ezio scales these buildings via canopies, platforms and signs and most interestingly, he climbs structures using bricks which poke out a slight amount. The engineers of these buildings were either very sloppy, or extremely helpful to those who could run from the law using a vertical stretch of wall.

Lets consider a pulley system set up in such a way:

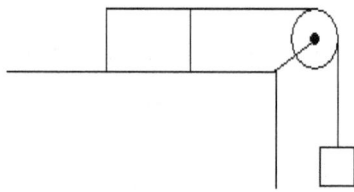

Where the left mass is m_1 and the right is m_2 and they are connected by a string on a frictionless pulley. If m_1 is the larger of the two masses there are two things that could happen depending on the conditions. First, if there is no friction on the surface m_1 is placed on, the force of gravity on m_2 will pull the entire system towards the ground. This occurs because of gravity forcing m_2 downward causing the whole system to move. The force of tension which was discussed before, does not deal with the motion of this mass. It is important to note that m_1 does have forces acting on it. There is the force of gravity and its normal force which are equal and opposite in direction, thus meaning it contributes no force for movement. Unlike the second mass, this mass is moving purely due to the force of tension provided by the string. So using the fact that F=ma we can determine what force is being applied on this system if we know the weight of m_2 .

A second scenario of this situation, and that would be the inclusion of friction between m_1 and the surface. Friction is a force which a opposes an object's motion. This introduces the coefficient of friction denoted by the Greek letter μ . Whether this is a static or kinetic friction means the symbols are denoted as μ_s and μ_k respectively. Static friction has to deal with getting a motionless object to move, while kinetic is the friction involved when an object is already in motion. It is important to note that for a given material, the coefficient of static friction will always be greater than kinetic. It is much easier to keep an object in motion, rather than put it in motion. Try pushing a heavy object along a floor and take note of which is more difficult: starting the motion or maintaining it. Much like we did before, we can create a sum of all the forces to determine what will occur in the pulley system if we introduce friction with m_1 (neglecting friction on the pulley).

$$\Sigma F = -\mu_s m_1 g - T_1 + T_2 + m_2 g$$

The first thing that needs to be noticed is the use of the negative and positive signs in the equation. Arbitrarily we can choose whether the system going right or left is positive. In this case we have chose right as positive. If we have a large mass for m_1 , and/or it has a high coefficient of static friction with respect to the forces provided by m_2 , this system will remain motionless. The opposite occurs if m_2 is very large in comparison to m_1 . This principal can be used to show what is happening when our hero of Assassin's Creed, Ezio, is scaling the side of a building on bricks alone.

In Ezio's defense, at least some of the bricks he climbs are poking out slightly from the sides of a building, but not much. When his hands are on a brick, the very same principal of the pulley system is applied to this situation, with minor differences. In the place of m_1 , we have Ezio's fingers and for m_2 we now have the mass of a whole person. We have to consider one new important difference: the replacement of the pulley with his palm and fingers clenching on the edge of the brick. This and friction are the only reasons he is able to hang off a brick supposedly. Judging by the character's height and abundance of weapons stored on him, we can estimate he weighs approximately 120kg. This would mean relating back to our pulley system, between the tiny force of friction between his hands and the brick and the slight force his palms exert to hold onto the edge. He would have to clench onto the ledge with more than 1276 N of force (Using F=ma). Unless Ezio has extremely strong hands, it is doubtful that he could up his weight on a single brick, much less scale a building by launching himself off these bricks. The same can be applied to any of these Sandbox games which allow the player to perform this action, it is just not very feasible.

It is important to note that just because this doesn't seem feasible, does not mean it is impossible for this to occur. The person climbing the building would just need a very large amount of strength in their fingers and maybe have some finger protectors which raise the coefficient of static friction between the fingers and the brick by a large amount. There is however a sandbox game which not only received tremendous praise and success like Assassin's Creed II, but its climbing mechanics are surprisingly accurate.

Shadow of the Colossus has the player take control of Wander who for reasons the story dictates, has him climbing disgustingly enormous beings called colossi. He has to slay these colossi with a sword that is the size of his forearm. Some of these colossi really do need the use of the word colossal as Wander, who appears to be around the height of 1.7m, is not even the size of the 'baby toe' of some of them. So we have a sword, the size of his forearm and he is the size of a colossal toe. We are not here to question why Wander is out of his mind or the logistics on why bringing something a colossi would not use as a toothpick as a weapon. We are here in the name of physics and climbing mechanics.

The reason why climbing is a chore for Wander should be immediately obvious given the difference in sizes. Each colossi throughout the game is thankfully covered with hair on important parts of their body along with conveniently placed platforms. Each colossi varies in size: some are truly colossal, others are rather small and are not entirely deserving of the title. Nonetheless, every situation involving climbing these beings follows a very reasonable, real world logic (assuming, of course, colossi are real). As was said, each colossi is covered with large strands of hair and platforms with ledges which someone with a lot of strength could grasp onto. Each colossi will try to forcefully dismount Wander as if it were a person disposing a fly. During these violent shakes, Wander has to cling on for dear life so he does not get sent crashing to the ground undoing a lot of progress and giving him further brain damage. At first glance, this does not seem any more plausible than Ezio scaling a building using nothing but bricks, but we are introduced to the saving grace called a grip gauge.

The game has limited the time Wander is able to hold onto a colossi as it shakes and twists in an attempt to send him soaring majestically to the ground below, thus introducing Wander as a physical being with physical limits. The same type of physics applies as in the case of Ezio climbing a building, since he has to rely on friction to help him stay on ledges and grasp hairs. Unlike Ezio, Wander has a limit and we know how strong he is. We cannot simply, as the player, go where we want for as long as we want, when we want. We have to meticulously plan camping spots where we can kneel on a platform, or the shoulders of a colossi to regain strength. Wander is an extremely powerful person, way above the average person; but he still has limits none the less.

As the game progresses the grip gauge (his strength), increases as well. This follows suit to a person is to lifting weights and running all of the time. He or she will have an increase in their strength and endurance as time progresses. The harder a person trains, the more results he or she will yield. Wander is in a similar situation. He is pushing his body to extreme limits for extended periods of time, then getting down times for his body to rest and build itself up. So although the idea of colossi is not plausible in our fleshy, carbon world, Wander and his ability to climb like very few is.

CENTRIPETAL MOTION – NOT A PARKING LOT!

Back in the humble beginnings, before the juggernaut series of Uncharted games and the emotionally gripping masterpiece that is The Last of Us, Naughty Dog created Crash Bandicoot. In the Naughty Dog years of Crash Bandicoot, four games were produced with this bandicoot that assumes human characteristics and wears pants. The first three games are wonderful platforming games, the fourth is a kart racing game, the latter being where our attention will be devoted.

Crash Team Racing (CTR) is certainly a gem and a game any kart racing fan should play. There have been many kart racing games throughout the years, but CTR truly stands out as one of the best. From the surprisingly wonderful introductory screens, to racing on the diverse and vibrant tracks in the simple yet compelling story mode, to powering up weapons with fruit to give the player the edge in the race, CTR is a well rounded gem. It is interesting to note the contrast in stories for Naughty Dog games. The Last of Us is an emotional roller coaster where key characters are killed, people are truly surviving by any means, it is a survival game through and through, a masterpiece, a literary opus. CTR follows the story of these heroes who have to win all the races in story mode so the Earth will not be made into a parking lot. So let's look at how an insane dog, devious scientist with a rocket in his head, sickeningly adorable polar bear and many other friends are able to drive these karts with concepts people take for granted while driving.

For those who know what this game is all about, or looked at gameplay videos online out of curiosity to provide context (hint, hint!), you'd be surprised to know (spoiler alert), that many of the mechanics in CTR obey the laws of physics. Some of the concepts are slightly exaggerated, but indeed accurate in many respects. We will look at three mechanics in order of increasing complexity to see why it is possible to have a giant tiger named Tiny can drive a kart the way he does. We obviously have to suspend disbelief a little, as we probably won't see an anthropomorphic (human like qualities) tiger driving a kart any time soon.

First up is the ability to bounce, leaving no wheels on the ground. This happens through the use of hydraulics, something used in a number of vehicles in real life. The bouncing typically happens one side of the car at a time as to prevent it from becoming airborne, many movies have displays of this. Rather than doing one side at a time, CTR has its racers bounce from all wheels at once. How do these hydraulics work?

We can think of hydraulics as a series of force tubes which press on the Earth's surface. If we recall the idea about every action having an equal and opposite reaction, that same idea is crucial here. The force tubes use some inner mechanics to press on the surface of the Earth (which apparently is difficult to move) and allows for a substantial enough force to launch the kart off the ground, much like how someone is able to jump (remember Mega Man and friends?). The hydraulics are coordinated in a such a way that the kart will becoming airborne without being lopsided. This means there needs to be some wonderful suspension and shock absorption to make for a less aggressive drive. Moving right along to mechanic number two.

We now are focusing on friction, the reason people are able to grip onto the road with their vehicles and take turns. We already have examined static and kinetic friction and how they allow, as well as, restrict movement. Without friction allowing vehicles to grip to the road, karts would be flying off the track, every which way and what have you. The important part of us bringing friction into our analysis is determining what type of friction we have between the road and tires; remembering that kinetic means moving and static means stationary. If you guessed that we are dealing with kinetic friction, you'd be wrong (unless you were thinking about someone slamming on their brakes and slid-ing on ice). Even though the tires are constantly moving, the same part of the tire is not constantly in contact with the road. At every moment while driving, new components of the tire are in contact with the road, thus, as far as the road is concerned (if roads could be concerned about things), the tire has a coefficient of static friction to be concerned with (yet again, assuming tires can be concerned with things; mine won't return my calls). Each instant a new component of the tire is touching the road, it appears to be stationary, hence the static friction. Now we will look at the most com-plication concept, the one you probably did not know was a concern (the title of the section gives it away): centripetal motion.

This is the part where things are going to get a little bit confusing as we explore how something can be con-stantly accelerating inward, but go in a uniform circular path. Imagine swinging a tennis ball tied to string in a cir-cular motion. When doing so, at any given point in time, there is an instantaneous velocity and a specific direction of travel. Regardless of what point in time we focus on the ball, it's always accelerating perpendicular to the velocity. In other words, the ball is always accelerating inward along the string, towards the person's hand spinning the ball. The velocity will always be at a 90 degree angle to this, so if the ball is cut loose at any time it would travel in that direction with no curve. Try doing this in a very safe place if it is con-fusing, as the explanation may be as clear as mud in the night. The most important and most confusing part of this is the acceleration inward.

We previously dealt with gravity, where the direction of acceleration pulls objects towards the surface of the Earth, not sideways. But if there was no inward acceleration, curves would not be possible, everything would have to travel in a designated line (note that we say line, not straight line, because a line is always straight). If we have an acceleration and a mass, this means that we have some type of force.

If we return to the case of the ball on a string, the reason this spinning is able to occur is due to the force of tension. Tension not being a factor would mean that someone has learned how to hover rocks in circular patterns, which would be incredible, but probably not practical. Based on the velocity, radius and mass, we can calculate the force of tension, or use the radius and velocity to calculate the acceleration alone. This is a very confusing topic, so take some time to absorb it, then come back to apply it to CTR.

In the case of a polar bear turning a corner in a kart, there are no strings, how does the inward acceleration occur? Why friction of course, that second mechanic was not there just as a food for thought, but as a foreshadowing. Friction is a force, and like all forces, has the crucial component of having an acceleration associated with it. When turning a corner, the velocity is tangent to the curve of the turn while there is an acceleration caused by friction occurring along the radius of the turn. Even though the driver will probably not complete a full circle, there is still a centre point where the driver turns around, so it is basically a pie shaped turn. The acceleration happens towards the corner of the pie slice and the velocity is following the crust tangentially. So if our heroes understand how pie-accelerations and force tubes work, they can save the Earth from being paved over.

MECHANICAL ADVANTAGE – PROPHET, SAVED BY PHYSICS

There are some people who after a majority of the bones in their body shattered, would probably give up hope and just accept fate. Others would be even more determined than they were before to ensure they get better as quickly as possible and not accept any excuses. Then there is one who does not have the choice about his critical condition and gets a suit and has to cope with the pain anyway. The last category is where the star of Crysis 2, Prophet, falls into. We are going to be examining Prophet and his nano-suit in the game Crysis 2.

Developed by Crytek and published by Electronic Arts, Crysis 2 features a soldier who unwillingly gains control of what the game calls a nanosuit which keeps this broken soldier together. On top of gaining the name Prophet, this soldier now has increased strength and physical resistance and the abilities to temporarily become invisible and conjure a barrier increasing the physical resistance even more. The structure of this suit even provides Prophet with enough support to be able to move as freely as he would without the life sustaining injuries. Can this form fitting suit really provide all of these advantages?

There is the possibility for the invisibility technology to work similar in principle to the way mirages occur. Mirages occur because light, after travelling in a colder air hits a packet of considerably hotter air which bends the light away from where it would travel without the hot air. Keep in mind the reason that people see is because light reflects off a surface, all the colours of the spectrum unseen to the eye are absorbed by the surface and the colour we see is what is not absorbed and reflected. Light bending when hitting hotter air is similar in principal to a vehicle going from a paved road to mud at an angle.

If a person is driving a vehicle along the road and decides to detour on a muddy side road, the tire that hits the mud first will be moving slower since there is less friction to allow the wheel grip to the mud causing slipping. The other wheel will still be on the road moving faster to the vehicle will start to pivot with the wheel in the mud as a pivotal point seeing as the wheel on the road is going faster. If the vehicle was going from the mud to the road at an angle, the vehicle would still pivot around the pivotal point of the wheel in the mud, but in the opposite direction. In the first case the vehicle would travel closer to being perpendicular to the road and mud meeting point(90 degrees, getting closer to going straight towards the mud). Whereas with the mud to road, the vehicle would pivot closer to travelling parallel to road and mud meeting point. Light acts the same way. Comparatively colder air is considered a slower medium like the mud and hot air acts light the road.

To emulate mirage conditions, the suit would become extremely hot in the immediate area so that light would bend around it. This would give the nanosuit that wavy, invisible look as one sees in a mirage. We would just have to assume in order to use this, the nanosuit compensates for the inevitable dehydration by absorbing water from the surrounding area and hydrating prophet. If invisibility cloaks were available for three dimensional objects, this would possibly be the technology they would follow in order to work.

The physical resistance would be completely plausible seeing as this nanosuit acts as a form fitting exoskeleton which is comprised of, most likely, a weaved metal alloy. If a person replaces blocking bullets and punches with their skin and bones with using a woven metal suit, clearly the resistance to physical damage is going to increase. This concept is nothing new seeing as knights and warriors did this many years ago (some people do it today as well); it is just that the technology is now better. A suit which keeps track of vitals and gives both physical strength and resistance without speed as a compromise is much better than a suit of metal which slows the user down and only increases resistance in certain areas since it is not possible to have the full body covered. Would it be possible to increase this resistance while in the line of action?

There is the possibility for an increase in resistance by having the nanosuit contract the woven structure so that the nanosuit becomes tighter and thus more resistive. This is where a compromise must come in. There needs to be a time limit as to how long Prophet is able to activate this 'barrier'. Since there is still a person inside the nanosuit, if the suit is form fitting prior to contraction, then the contraction will result in Prophet being crushed. A human body can only withstand so much pain and punishment before going into shock or experiencing other serious repercussions. So putting a limit to how long this is activated is a necessary seeing compromise as Prophet is already in critical condition. Now we have our last area to focus on: the increase in strength.

In order for us to determine whether the strength increase is plausible or not, we need to examine mechanical advantage involving lever systems. There are three types of simple lever systems which can be set up. The first is something seen in classic cartoons where someone will try to move a big boulder using a plank of wood and a rock. An end of the plank of wood is placed under the boulder, the other end is placed in the hands of the user and somewhere in between there is a rock under the plank. Whoever is trying to move this boulder will push down on the plank of wood, the rock will act as a point to pivot around (known as a fulcrum point), which will allow the other end of the plank push up on the boulder and potentially move it. The longer the board is, the easier it will be to move the boulder as it creates more leverage. This is assuming the length of the board is greater on the side of the applied force than on the side of the output. In other words, greater the length of what is known as the force arm, the easier it becomes to move the object in question.

The second type of simple lever is what we see when using a wheelbarrow. This time a person is lifting the wheelbarrow off of the ground by the handles at the end. There is a fulcrum point at the opposite end, the wheel in this case, and an output force in between which would be the lifting of the container of the wheelbarrow. This is another example of lifting becoming easier as the force arm where the force is applied is a longer distance to the fulcrum point than the distance from the fulcrum point to the output.

The final type of simple lever is observed when someone is simply lifting a weight with their arm. This time the force arm is the forearm to the connected fulcrum point (the elbow) and the output force is placed in the hand. This means unlike the other two situations, the input force is less than the output force thus creating a disadvantage. Regardless of which lever is being used they all follow the same formula which has been eluded to:

$$Mechanical\ Advantage = \frac{(Output\ Force)}{(Input\ Force)} \equiv \frac{(Input\ Force\ Arm)}{(Output\ Force\ Arm)}$$

This equation is applicable keeping in mind that this is an ideal situation without other forces to consider such as friction and elasticity.

With the first two types of levers increasing the mechanical advantage is a lot easier seeing as we just need to increase the length of the force arm. Returning to the plank of wood and the boulder, assuming the wood does not break, if we continue to increase the length of the wood on the input side of the fulcrum point the output force will require less effort. Given a big enough lever, anything can be moved by a person. Although it is not practical after a certain length, the same occurs if we increase the length of the handles on a wheelbarrow. So the mechanical advantage ratio will result in a value greater than one because the top value will greater than the bottom in the fraction (assuming the situations are arranged in the ways described before). If the ratio gives a mechanical advantage of one, this means that the input and output force are the same thus giving no advantage or disadvantage. The third type of lever gets a little bit more complicated though.

Returning to the example of someone's arm, the stronger the muscles in the forearm, the less of a disadvantage there would be. A stronger muscle does not give a greater advantage due to the ability to supply more of an input force though. Mechanical advantage is a ratio and a greater input force would get a greater output force, but the ratio would remain the same because they increase proportionally to one another in an ideal case. More of an advantage will be gained when the forearm becomes extended over a greater area, which would mean a larger input force arm. Regardless of the strength, the ratio of these will always result in a value less than one, resulting in constant mechanical disadvantage. There is a way for the nanosuit to at least help tend towards an advantage.

As with a person lifting a weight, the mechanical disadvantage is always present, but at least there is mobility involved with that disadvantage. The other two levers have a limited range of motion compared to the third lever. With the nanosuit acting as an exoskeleton, Prophet has the advantage of mobility as well as a mechanical advantage very close to one. The nanosuit having its woven steel and acting as an exterior muscle will add to the strength. This is what we were initially analyzing, but we have the chance to give Prophet more potential to lift objects. Keeping in mind the nanosuit acts as exterior muscles, the suit is form fitting all over the body, so we have a muscle extension across the whole arm. This means that the input force can be used at the hand rather than somewhere further back. If we have the force being applied at the hand where the weight is being lifted, the force arms will be equal in length resulting in neither a mechanical advantage or disadvantage. So not only will Prophet gain strength from the nanosuit, but he will reduce his disadvantage giving him true super strength. This is one amazing nanosuit!

SECTION II - ENERGY
ELASTICITY – A ROLENTOLESS PURSUIT FOR THE TRUTH

Thanks to RetrowareTVs General Brent of Gaming Legends for his half suggestion. He suggested I cover something in a Mega Man X game, so I am going to in honour of the Blue Bomber's 25th anniversary present. Without further ado, let us jump right into Mega Man X Street Fighter (hence the half suggestion, really not the same as X6, at least we briefly covered the first X game).

I actually enjoyed this game quite a bit. Other than my randomly appearing, impossible to get E-tanks, I am not going to complain about my free game (seriously, it's free, go get it!). It is quite fun and the creators should be proud. We are here for science! Let us move right to what is probably my favourite subtle reference in this game: Rolento jumping on a pole much like a certain someone from the beloved NES Duck Tales (Thanks to Pat the NES Punk for inspiring me to include this with his rendition of the theme song).

Rolento and Scrooge McDuck both are able to bounce on their respective pole and cane, but there is a little difference between the way each bounces. Scrooge's cane resembles a pogo the way it bounces since it is a straight up and down motion. Whereas Rolento's adds a side to side motion in the mix. Let us examine the former before moving onto the more complex wobbly pole.

Scrooge McDuck's cane, as mentioned, acts very much like a strong pogo stick. How it works is that there is a spring inside of the cane which has a large enough elastic potential to allow a duck to bounce at such great heights. It follows this equation:

$$U_E = \frac{1}{2}kx^2$$

Where U_E is the elastic potential energy, k is the spring constant (dictates how powerful a spring it is, its springiness) and x is the amount of displacement from the spring's uncompressed state. If uncompressed, there is no elastic potential energy to be used, because if x is zero, then the whole equation is zero. The more the spring is compressed, the more potential energy it offers and can be converted into kinetic energy upon decompression. The greater the spring constant, the less the spring needs to be compressed to create such great amounts of kinetic energy. In our case, the kinetic energy is used to launch a duck into the air. This equation only covers motion in one dimension though. What about in the case of Rolento? His pole actually bends quite a considerable amount, so how is he able to maintain so much control?

There are flexible poles, much like you see in Olympic pole vaulting. They are made of either carbon fibre or fibreglass. It can be seen that those poles offer a very large amount of resistance and maintain durability. But when pole vaulter's engage in their events, they use a considerable amount of strength and careful positioning to place their poles in a very specific spot in order to bend in the fashion they do. Even at a shorter length vaulter's could not stick this into the ground at any point and have it stay in one spot and support weight at any jumping direction. I do not advise getting a flexible pole and trying to jump on it, chances are something/someone will get hurt and/or broken (trust me). So the design of this pole has to be some sort of a hybrid design between a pogo and a pole vaulter's pole.

In the game we can see this as one solid stick, so any of the complicated design is left to internal design. Firstly there needs to be a weight in the bottom of the pole to allow for a stable bounce despite the sporadic movement. There could possibly be a large amount of lead, which is a very dense and heavy metal. The lead being so heavy would have much more of an affinity with the ground and would be rather stable on the ground if large enough. Along with this, there could be some sort of a suction cup system on the bottom. Secondly to allow for such a wavy movement at the bottom of the pole, the walls needs to be composed of a very thin fibreglass material. To compensate for the lacking durability there could be springs to guide movement to particular limits as well as provide much needed stability. Not so fun fact: All springs have what is called an elastic limit. If springs are stretched beyond this limit they will never retain their original shape and form as the metal inside is now damaged and can only be fixed with reforming the metal under intense heat. You know how you killed your slinky as a kid? That is why. Returning to the pole design, the top part could be completely solid fibreglass to add strength but not a lot of weight considering the bottom has lead, that is more than enough to lift.

Keep in mind this is just theorizing how the engineering of this pole would allow Rolento to be able to pull off all of those fantastic moves. My lead supply has been rather low as of late and my fibreglass pole creator has been in the shop for weeks. I would love to not only make this myself, but test it as well. If you manage to make yours work, please let me know, that would be awesome. Now that we introduced an equation for elastic potential energy, we will try understand it more by applying it to a strange

Japanese game most people reading this have never heard of, Umihara Kawase.

Umihara Kawase, developed by TNN and published by someone (I seriously can't find out who), is like a few games mentioned already, a platforming game with a little twist. One twist in this game that makes it so unique is the use of a strange bungee-cord fishing rod (going to just refer to it is a rod for simplicity). It is important to note that using a hook to latch onto a ledge, it is possible to lift a little girl with this rod if it is possible to pull down a helicopter with a harpoon. We are not going to examine the visuals, goal, or death animation, just accept that the game is rather odd and analyze the rod mechanics and determine a spring constant; thanks to Ben Hall of Video Game Take-Out for suggesting we cover it.

The rod allows for the main character, Umihara Kawase, to be able to latch onto hard to reach ledges, latch on and pull herself by take advantage of potential energies, elasticity being one of them. Even though elastic potential energy is one of the energies used to accomplish this feat, it is not the only one and this is where we have to apply the idea of "energy can be neither created nor destroyed, just converted". For the purpose of this explanation, we will keep things simple and ignore any potential losses due to friction, heat, and others and focus on the three primary energies we need to concern ourselves with: elastic, gravitational and kinematic.

We have already examined elastic potential energy, this is provided by elastic objects that are either compressed or expanded. In the case of Umihara Kawase, we are going to be worrying about extensions.

Gravitational potential energy (we will use Earth as a

reference) is provided when an object is lifted from the surface of the Earth, or something on the the surface of the Earth (in the air). Gravitational potential energy it has as it is based on three factors:

$$U_g = mgh$$

Where h is the distance between the surface and the respective object, m is the mass and g is the acceleration due to gravity. This means that gravitational potential energy will increase as mass and distance from the surface do.

Finally we have kinematic energy, which is the energy an object possesses while in motion. This is based on two factors:

$$U_k = \frac{1}{2}mv^2$$

Where m is the the mass and v is the velocity, both terms we are already familiar with. Now we need to combine these three energies into one equation:

$$U_1 = U_2$$

Where U_1 and U_2 are a combination of energies at two different reference points. This maintains consistency with the notion of energy neither being created nor destroyed, this will be easier to see once we apply some energies to this equality.

In order to set up our potential solution, we need to pick two points in time to examine keeping in mind that energy is converted. Our first reference point is will be when Umihara Kawase is hanging as low as her rod will go, but not moving. The second will be the after she detaches herself and is still moving, almost about to touch the ground. We will write the general formula involving all of the energies we are concerned with, then begin elimination of parts based on the conditions we set up:

$$U_{k_1}+U_{g_1}+U_{E_1}=U_{k_2}+U_{g_2}+U_{E_2}$$

Based on our conditions, we can already elimante three variables to make our lives much easier. U_{k_1} can be eliminated because we applied the condition that Umihara Kawase (how many times has this name been said and I still don't know how to pronounce it) is not moving when her rod is extended and she is hanging. U_{E_2} and U_{g_2} can also be eliminated because the rod is detached and no longer providing its elastic potential energy and we chose a moment where she still has not quite hit the ground, thus the height is negligible. We are now reduced to:

$$U_{g_1}+U_{E_1}=U_{k_2}\rightarrow mg\,h_1+\frac{1}{2}k\,x_1^2=\frac{1}{2}mv_2^2$$

We can arrange this to solve for our spring constant:

$$k=\frac{mv_2^2-2mg\,h_1}{x_1^2}$$

All that has to be done is substitute in some values and solve. Keep in mind that this game may not take place on Earth, but for simplicity we could assume the same gravity if we wish. The mass of Umihara Kawase is a matter of opinion as well. Thus I advocate you to pick some numbers yourself and come up with a value for the spring constant. This could be done by watching a gameplay video online to determine your own reference points and come up with a value. I would do it for you here, but I still can't pronounce Umihara Kawase. Doing calculations on a video game character I can't pronounce the name of is the equivalent of taking someone you don't know the name of on a date, it just isn't for me.

IDEAL GAS LAW – SPRING BREEZE AWAY INTO THE NIGHTS

This time our adventure turns our focus towards a little pink blob which has unusual physique, cloning and flight abilities; we are now focusing on Hal Laboratory and Nintendo's, Kirby. Kirby is a pink ball with little feet, flap-like arms, cute oval eyes and a mouth which is small when closed and massive when opened. There is not much of a structure to Kirby. If we were to put shoes, flaps and draw a face on an exercise ball we would have him designed perfectly. Kirby has an incredibly unusual combat tactic. He is able to open his fat mouth and suck up an enemy to clone their abilities. This is limited to certain types of enemies. If he finds a mushroom or a sweeping witch, he will not clone their abilities, but if Kirby finds an enemy flying around with a jet pack who can charge and boost, or a sword knight, he can suck them up and miraculously get their abilities to slash and dash. Kirby is limited to cloning one ability at a time so he is not able to become a sword wielding, laser shooting, jet boosting chef of a pink blob; which is quite unfortunate due to how fun that sounds. We will examine this fun stuff stuff later. For the time being, we will focus our attention on Kirby's unique flying abilities.

When Kirby decides to open his fat yap to suck in air and not sacrifice an unwilling enemy, he can grow to about double his size. With this excess amount of air in his mouth, Kirby is able to jump and slowly glide through the air and if need be, flap his arms to fly indefinitely. When relaxed Kirby jumps, he falls to the ground by the force of gravity and does not slowly float down, yet he can fly. Birds are able to fly due to their aerodynamic wings and weight (remember?), but Kirby does not have aerodynamic wings and we can not examine his weight. Let us examine this closer.

We discussed before how our fearless soldiers of Contra could not change directions while jumping, nor could they double jump due to a lack of pressure beneath them. Kirby on the other hand, does not encounter the same problem that our soldiers do. Kirby seemingly has a considerable weight difference than alien destroyers since he resembles a walking balloon more than a person. So if Kirby resembles a balloon, then when he makes a flapping motion, he does not need nearly as much built up pressure to allow him to propel upwards. We can not decide whether this is a possibility or not based on pure speculation. We need to analyze this, so we need to introduce the idea of the ideal gas law and other concepts.

There are many different ways to classify the characteristic of matter: weight, strength, colour, odor, most importantly for us, the state of matter. There are three commons states in which matter can be classified: solid, liquid and gas. A solid is something like a book, spoon or computer. A solid has a defined shape and will not take on the shape of a container it is put it. It can be malleable and does not have to have a permanent shape, but it certainly can not take the shape of a container. Molecules in a solid are tightly bound to one another, which gives most solids the strength they have. A liquid will flow, the rate will depend on how thick (viscosity) the liquid is and it will take the shape of any container. Molecules will move around inside of the liquid and not be so rigidly bound to one another. The way liquids occupy a container is determined by gravity as they will fill from the bottom upwards. Finally we have gases, which also take on the shape of any container but do so in a different manner than liquids. Gases will expand and spread out to reduce pressure and will attempt to expand their container if no escape routes are available.

If no escape routes for air are available in a container, we consider it a closed system. Closed systems have a wonderful advantage for analyzing gases because then certain characteristics become constant and we can at least see the cause and effect of certain situations. Take for example, say we have air in a balloon, we know if the balloon is a closed system, then the same amount of particles will be in the balloon at all times. With the same situation, if we raise the temperature of the air molecules inside the balloon, several situations can occur: pressure remains constant or while the volume expands, volume remains constant and the pressure increases, a combination of the former or the balloon pops and we have nothing to analyze since we destroyed our experimental equipment. We can come to the conclusions we did by various methods, the first being Kinetic Molecular Theory (KMT).

KMT deals with the movement of particles, for our case, inside a container. Every gas molecule which moves inside of our balloon has a certain amount of energy. Each molecule will be moving in different directions at different speeds and energies. Some will collide with other molecules, some with the container walls and some will just travel pain free. The collision of molecules with the container walls dictates our pressure. The different units we could use to analyze pressure such as the atmosphere, bar or pound per square inch; but keeping with our trend of using SI units (International System of Units; the SI abbreviation is used in every language but is the abbreviation for the French name), we will use the pascal. A pascal (Pa) is measured in newtons per metre squared, which refers to how much force is spread across an square area of one metre.

Returning to our case of the balloon, if we increase the temperature, KMT dictates the the air molecules inside will gain energy, which will cause them to move faster and increase the pressure. The pressure could balance itself out and return to its original value if the balloon expands an appropriate amount. There would be a point where the pressure and volume are increasing to the point where the volume can increase no more as the balloon has a limit to its volume. This leads to our case of the pressure increasing while our volume remains constant, followed swiftly by an exploding balloon in the face of perpetrator. There is a law which deals with the balance of the temperature, pressure and volume called: The Ideal Gas Law.

The ideal gas law can be arrived at using the KMT and some assumptions to arrive at the following equation:

$$pV = nRT$$

Where p is the pressure, V is volume, n is the number of moles (amount of gas), R is the universal gas constant and T is the temperature. Pressure is measured in pascals and volume in metres cubed. The other side of the equation introduces new concepts though. The number of moles is straight forward as it chops two characters off the end to become a mol which saves people precious fractions of a second every time they would have had to write the whole word. The universal gas constant has the value of 8.314 J/K*mol which introduces two new units. First we have a joule (J), which is a newton metre, the amount of force spread over a distance of one metre. Secondly we have kelvin (K), which is what not only the unit we use for temperature in the ideal gas law, but a unit of temperature in the gas constant as well. The idea of the SI unit kelvin has an interesting place in science.

When measuring temperature whether it be on a thermometer outside or by a person on the news predicting the weather, they use the units of Celsius or Fahrenheit depending on the location. The unit kelvin however is used widely in science for two reasons: it is the SI unit for temperature and it is used as a basis for absolute zero. Zero kelvin is when absolute zero occurs, which has a serious implication: if the temperature is at absolute zero, all motion stops. Every gas molecule, old lady crossing the street and espresso coming from a machine completely stop. This is an incredibly low temperature though: -273.15 degrees Celsius. Although some Canadian winters feel like that, we have a long way to go to get to that temperature. As a reference point to determine the temperature in kelvin, we need to know that zero degrees Celsius is equal to 273.15 kelvin. Now that we have all of this established, we can return to discussing Kirby.

Kirby can follow the ideal gas law perfectly as a non-rigid container which allows for pressure, volume and temperature variations. We first need to make a couple of assumptions in order to determine the volume of Kirby. Kirby appears in a huge cross-over fighting game published by Nintendo called Super Smash Bros. Brawl. This gives us a great advantage for our analysis as we can now determine his height as he is now standing beside characters from many other games, including Samus, for whom we already made a height assumption. In the game Kirby appears to be just short of half the height of Samus, so we will say he is 0.80m tall when in non-devouring mode. Our next assumption is that the insides of Kirby are like a balloon with a negligible thickness. Since Kirby is clearly not a form of life we have on Earth, it is safe to assume his body is made with an unfamiliar substance as well which allows for structure as well as expansion.

When Kirby is flying around the Spring Breeze Woods, we will assume ideal conditions for a nice day to go out for a walk of 20 degrees Celsius and 101.3kPa of pressure. These values are actually considered ideal conditions to some when conducting experiments and doing sample calculations with the ideal gas law. This is not an unreasonable temperature for a nice spring day, so it will work out perfectly for our application of the ideal gas law. We need to make one last assumption before doing some calculations: the air has to be dry so we are not introducing unwanted water molecules in the air and can use the average composition of air at the previously stated conditions. Using the average composition of air which is mostly comprised of nitrogen and oxygen, we can access their respective masses on the periodic table, multiply the value by the percentage it composes and add all the values together. Doing the math, or simply accessing a table, we can get that the molar mass of dry air is 28.965g/mol. Now we are ready to do our pre-calculations for the weight of air inside of Kirby.

First we are going to convert our universal gas constant into different SI units since he know we are looking for a weight in the end, not the number of moles. We do this by first converting our mass of dry air into kilograms recalling the prefix kilo means a thousand times the base measurement: (28.965g/mol)/(1000g/kg) = 0.028965kg/mol. Next we can divide the gas constant by our mass of dry air as follows: (8.314J/K*mol)/(0.028965kg/mol) = 287.04J/K*kg. Finally we need to calculate the volume of Kirby in his non-puffed up form and when he is flying. We assumed a height of 0.80m relaxed and double that value when full of air. Using the formula for volume of the sphere we get the volumes in the following way, recalling the height is a diametre, so we talk half that value for the radius:

$$\frac{4}{3}\pi r^3 = \frac{4}{3}\pi(0.40m)^3 = 0.268m^3 \text{ and } \frac{4}{3}\pi(0.80m)^3 = 2.15m^3$$

We are now ready to calculate the weight of the air inside our cute, pink blob. First we will calculate the air inside of Kirby in his relaxed state:

$$pV = mRT \rightarrow m = \frac{pV}{RT} = \frac{[(101300Pa)0.268m^3]}{[(287.04J/K*kg)293.15K]} = 0.32kg$$

In a similar fashion we can find weight of the puffed up form:

$$m = \frac{pV}{RT} = \frac{[(101300Pa)2.15m^3]}{[(287.04J/K*kg)293.15K]} = 2.6kg$$

Keep in mind that we could have used a different pressure or temperature for the calculation and the weights would remain the same. We have a closed system inside of Kirby and the equation would essentially act as a balancing act. If we raise the temperature, since the volume is rigid, the pressure would increase to balance out the increase. The opposite would occur when lowering either value. Now that we have those values, we can begin to examine whether Kirby can float or not. First let us return to the case of a balloon.

There are different factors to consider when finding out why a balloon floats. The first is what the balloon is full of. If a balloon at a surprise party (which if he or she is like me, they planned it themselves and is the only one there) is full of helium, it will definitely float. Helium is much lighter than the overall composition of air, so the heavier components will lower, while the lighter rise. This is known as buoyancy. If we think of a buoy floating on water, the buoy's weight exerts less pressure than the water it is on. So the pressure of the water will continue to let the buoy float. This is why at a lake it is possible to see light drift wood floating, while something like a rock will sink to the bottom. This brings up an interesting segue.

A person standing at the side of a lake with a nice, flat rock is able to throw the rock fast and at a small angle with respect to the horizontal surface to skip it along the surface. This happens because the faster something hits the surface of water, the more water acts like a solid. This is why someone jumping off a diving board onto their stomach can bring forth a tremendous amount of pain. The water offers more resistance, or more pressure, so the rock is able to bounce off the surface similar to why a buoy can float. When the rock strikes the surface of the water each time it loses a lot of momentum and transfers energy to the water so the rock will become less likely to bounce after each contact with the surface. This will only happen if the rock is thrown fast enough and at an angle. Throwing a rock straight down will not allow for skipping along the surface, just a quicker sinking of the rock. This is why we can not skip monkeys.

That comment is not nearly as out of place as one might think. Rare and Nintendo developed and released a platforming game called Donkey Kong Country 3: Dixie Kong's Double Trouble which allowed the player to take control of two prime apes named Dixie Kong and Kiddy Kong. At certain points in the game the player is able to assume the control of Kiddy and let him run really fast, roll off of a ledge, and bounce off the surface of the water three times and continue to bounce higher with each skip. Although this is a lot of fun, it is clearly way off base with any scientific merit. The first major problem is the sheer size of a gorilla. It is difficult to skip smaller rocks and get the pressure of the water to allow the rock to skip, much less greatly increasing the size and surface area of that rock to a gorilla. The next concern is the water acting as a substance to offer elasticity like a trampoline, rather than something to promote energy loss. Although very fun, a trampoline-water-gorilla makes no sense at all.

Where buoyancy does work is with an adorable and heart-warming puzzle-platforming game developed by Game Arts and published by GungHo Online Entertainment called, Dokuro. Dokuro has players take control of a prince named Dokuro and allows them to use a variety of techniques to solve puzzles using the PlayStation Vita's unique control scheme. The ability demanding our attention is the one letting the player control water levels by drawing on the screen and using the buoyancy of wooden boxes to solve puzzles. Using lighter wood, like a drift wood, and possibly a container with light gases as the bottom, a floating crate is very achievable.

Enough sidetracking, let's return to the example of the balloon. If we fill it up with air rather than helium we see different results in the way the balloon floats. The more we fill it up with air, the more buoyant it will be in the air, but it will always continue to fall slowly towards the ground despite how much we fill it. The more we fill the balloon with air, there will be less of a pressure difference between the air inside and outside of the balloon, but ultimately the weight of the balloon will always bring it towards the ground. The other smaller factor to consider is the weight of the gases a person breathes out in order to fill up a balloon. When we breathe air in, the composition is different when breathing out because our body uses up oxygen and will dispose of unwanted gases. This would have to be a considerably large weight difference to compensate for the weight of the balloon though. Knowing this, is Kirby still able to fly?

We already established that Kirby is composed of some type of material we are not familiar with which is durable and has a lot of elasticity, while on that notion we will assume he is light as well. This is not unreasonable as we have no idea what powers and compounds the universe holds millions of light years away. So we will be treating Kirby as a balloon essentially. When he has very little air inside of him, his relaxed state weighs him down enough so he falls at the rate any object would under the force of gravity. In the relaxed state being full of 0.32kg of air, clearly the weight of Kirby is too great to allow for the pressures inside and out to be close to one another. In other words, uninflated Kirby does not allow for any buoyancy. When he is full of air, we reduce the difference in pressures between the inside of Kirby and the surrounding air, so he should be able to float like a balloon since the weight of air inside is a whopping 2.6kg. This also allows Kirby to be capable of flying through the air flapping his little arms as the faster he flaps them, the more pressure he can exert on the air beneath them to boost him up. Since Kirby can float, the extra push his arms give allows him to fly so long as he remains blown up and as a closed system. So unlike our fleshy, humankind, Kirby is actually able to float, fly and change directions in mid air while still obeying physics. The heroes and villains of all other platforming games must be jealous of this cute puff ball and his unique abilities.

As a side note, there is a weird character which appears later in the Zelda series by the name of Tingle which typically is floating around with a balloon attached to his back. Tingle is another character which obeys the laws of physics when it comes to filling objects with gases. Take for example in the game The Legend of Zelda: Oracle of Seasons; there is never a claim to what Tingle actually puts in the balloon. It could be hydrogen for all we know, which is extremely light and could easily lift him in the air seeing as the balloon is much larger than he is. Filling anything with hydrogen, however, is extremely dangerous because it is a substance which is really explosive, especially in compressed containers. This is why cars are not fueled by it, or blimps no longer contain it.

While on the subject of balloon-like creatures, this seems like the most appropriate time to talk about a caveman that uses a little bit of bouncing to toss enemies to soar majestically through the air. Thanks to Brent Blauser (General Brent) of *Gaming* Legends for the following suggestion.

Tomba! is a side-scrolling platforming game that was one of two games developed by Whoopee Camp (the second being Tomba 2). The game stood out and still does today as a unique platforming game because the player is immersed in a rich and vibrant world with its beautiful graphics, where he or she is side-scrolling in different planes. What that means is that even though the player is typically confined to moving left and right, there are particular locations where there is the possibility to move to a different sort of track where the player is able to run parallel to the one previously explored. It gives the game a very strange feel as it is a hybrid between what would be classified as two and three dimensional platforming games. Outside of this hybrid nature of the game, the aforementioned bouncing to enemy toss is of particular interest to focus on.

If someone wanted to throw something, a rationale person would simply stand on the ground, pick up the item and give it a toss. Tomba however, is not exactly rationale as he missed the memo of all the people who hurt themselves jumping onto exercise balls (I swear I never tried this). Tomba jumps onto his enemies, bounces up a bit, does a flip in the air and tosses them. Lets go step by step and examine how he is capable of doing so.

The first part is easy, Tomba has what is considered to be 'floaty' jumping, this returns to the idea of having lower gravitational pull as suggested while examining Contra. The lower gravitational pull not only makes jumping easier on the behalf of Tomba, but it will be slower because there is not as much force to push him down rapidly.

Next we have Tomba grasping his foe and bouncing up slightly with the enemy (sometimes inanimate objects, but we'll just say enemies for simplicity) in hand. In the case of enemies which are stationed on the ground, they typically seem to have slightly bouncy qualities, using only the most technical of terminology. With this, Tomba is able to press down to utilize some of that wonderful elastic potential energy to convert into kinetic energy. Along with the elasticity being taken advantage of Tomba has the ability to push himself upwards with a decent amount of force seeing as he is on relatively stable surfaces, stable enough to support his weight anyways. The combination of elasticity, pushing and lower gravitational pull would allow Tomba to be able to launch himself in the air with a decent amount of height as depicted in the game. Now onto the flip-toss.

The flip-toss is where things start to get a little more complicate. If Tomba was jumping without an enemy in hand, this delayed front flip is certainly possible. People are able to do front flips that did not involve jumping with a particular direction of momentum, by quickly jerking parts of their body to shift their weight and do a flip. This is no easy task, but is definitely possible; if you are trained to properly dive or do stunts on a trampoline, give it a go, it is very fun. To do so with an enemy in hand which possibly weighs quite a bit just is not possible though. However, if the enemies resemble Kirby in any respect, light on the inside, strong elastic walls and are basically like living balloons, then Tomba could surely pull off this fantastic maneuver. Maybe Tomba not only puts an end to the evil Koma, but he obeys physics as well!

Lets show another way these ideas presented with Kirby can be applied by looking at the suggestion presented by Michael Anthony of *Still Loading* at RetrowareTV: Nights Into Dreams.

For those who are not familiar with the game, the bulk of the gameplay lets the player assume the role of the character Nights in a third person flight game. There are many abilities to analyze, but we will focus on the core gameplay mechanic: someone without wings being able to fly around so gracefully (we're approaching flight in an all new direction now, no more gravitons for us!). There are many ways we can approach this issue, each with its own set of assumptions. We are going to look at whether this is possible or not using thermodynamics. First we need to have a brief understanding of how matter reacts due to changes in temperature.

All matter (matter referring to any physical substance found in our universe) will react to a change in temperature. For example, if temperature decreases enough a liquid will turn into a solid. This can be seen by putting water on a tray, placing it in the freezer and eventually receiving ice (solid water). This is because as the temperature decreases, the molecules the water is composed of slow down and form more rigid bonds. It is obvious when water turns solid when it no longer takes the shape of containers unlike liquid water. Another example, although usually done using pressure rather than a temperature change, gas can turn into a solid like when carbon dioxide turns into dry ice when introduced to sufficient pressure. The change can be obtained with a decrease in temperature as well, but it is not nearly as cost effective as the prior method. This latter method is more important for us to consider for Nights though.

During flying sequences, Nights is able to go up, down, left, right and everything in between in order to swoop through checkpoints and dodge terrain. Although there is tight control over the character, Nights is only able to make movements in relatively gradual changes. The paths Nights is able to move along are similar to those plastic tracks used to race toy cars. Although there can be winding and twisting, the path cannot switch rapidly like a V-shape. To justify how Nights soars through the air, we will return to turning gases into a solid and a very important assumption.

We are going to assume that Nights is an extremely cold individual, so cold that a powerful exhale can cause the surrounding air to change states to a solid. The reason that Nights does not freeze the surrounding atmosphere at all times is due to its (Nights has no defined sex) clothing being a perpetually warm insulator. As mentioned before, we are going to assume that the breath of Nights is able to create solid air; enough solid air for it to essentially act like a slide. Air rigid and powerful enough to support the weight of Nights 'soaring' through the air.

Seeing as Nights is making an air-slide, this explains why the changes in motion have to be gradual. The breath can only be blown in the general direction of motion to maintain forward momentum, much like with a toy car racing track. If there is a V-shape in a toy track, the car will not be able to continue. Similarly Nights will not be able to continue either, but rather than hit a toy track, Nights will begin plummeting in a free fall pending no intervention is made. But would Nights be able soar in a loop like in the game without hitting the previously created slide?

Surprisingly enough it makes sense that Nights is able to complete a full loop to return to the same point and not hitting an air-slide. The reason being is that the slides would be temporary because the surrounding air in the atmosphere would certainly return the air close to its original temperature and state. Much like how dry ice rapidly changes to gas outside of a thermally insulating container. With the assumption of being unbelievably cold, Nights surely can fly.

The assumptions made are very unlikely to apply to real life, but they are fun assumptions to make. It always creates an interesting situation, testing limits and trying to make possible of the impossible. Maybe this situation is possible, anything can happen in a dream.

ACTIVATION ENERGY – BETTER HURRY!

Pressing on with some more thermodynamics. We are now going to observe activation energy done incredibly well with an unusual multiplayer game: Atomic Bomberman. Developed and published by Interplay, Atomic Bomberman was a game that had similar concepts to most of the other games in the series, but still sticks out like a sore thumb with its unusual animation and narration. We could examine more recent games which are still available to purchase or older ones which are not. Either way this analysis will follow the same principles as most games in the series follow the same premise: use bombs to blow up other players.

Atomic Bomberman has the player set up in a rectangular arena with other bombermen with a variety of obstacles in between the player and her or his target. There is always some type of pattern with indestructible and destructible blocks in between. In order to progress to the other bombermen controlled by other people or computer players, the player must plant bombs to destroy the destructible objects. This destruction can yield one of two results: getting one step closer to blowing up opposing players and the possibility of an item drop. These items drops can help the player grasp victory by upgrading the range of their explosion, increase the number of possible bombs to place, gain sliding bombs and bouncing bombs. The items drops can also hinder the player by making them drop their supply of bombs without permission or reverse the controls of the player, thus disorientating them. Victory conditions are simple: vanquish all other players. With the premise out of the way, we can now determine what is so accurate about this game and its bombs.

There are two types of reactions which occur in thermodynamics: endothermic and exothermic. The former involves the absorption of energy in order to change the state of the reactants. Reactants are whatever elements or compounds which are involved before a reaction occurs. An endothermic reaction is as simple as melting ice as we are adding energy into the system and it is absorbing it. Another example is baking a loaf of bread. We have an input of energy which is retained in the dough to create a loaf of bread in the end. Thus meaning in the end of an endothermic reaction, the total energy of the system will be greater in the end than the beginning.

An exothermic reaction is the opposite: the total energy of the system will be greater before the reaction than after. So while the reaction occurs energy will be given off. This could be in the form of light, heat or other forms of energy. An example is the burning of fossil fuels: as they burn and release many products in the forms of heat and various particles used to power the vehicle. Another example, which is the whole reason to introduce this concept, is the explosion of a bomb. The reactants inside a bomb have a much greater energy than the products which have been displaced in all sorts of directions along with the energy being released as combustion, light, heat and shrapnel.

Whether there is an exothermic or endothermic reaction, both of them require some sort of activation energy. In the case of a bomb, the activation energy would be the lighting of a fuse as the fire would provide the energy needed for combustion of the bomb to take place. This is why a bomb is safe when being held of stored, a great amount of input of energy is required for the reaction inside the bomb to take place.

As a segue, the contents inside an explosive are incredibly dangerous outside of their container. Say we are considering a bomb, the explosive compound inside is incredibly unstable and has an extremely low activation energy. Even the friction of the compound rubbing along a surface could create an explosion. This is due to the considerably lower activation energy. This is why the explosive compound is placed in a container; the reaction of the explosive and the inside of the container raises the activation energy to make it much more stable.

A very important aspect we also have to concern ourselves is enthalpy: the difference in energy before and after a reaction takes place ($\Delta H = H_{(final)} - H_{(initial)}$). The activation energy is not included in this measurement. It is just the initial and final states. So going back to the ice cube melting, the enthalpy would be a positive value as the final energy of the system is greater than the initial. In the case of the bomb, the enthalpy would be negative seeing as there is a greater amount of energy in the system as a bomb then blown up into pieces all over the place. We have to keep in mind that energy is neither created nor destroyed; the energy is simply transferred. So now that we know the bomb is accurate, let us see if the power ups are plausible.

The item pick ups can be thought of more as capability pick ups rather than actual power ups. Take for example if a bomberman picks up an item which allows for larger explosions. He now will be able to use bombs with larger amounts of explosive material inside. If he picks up an item which allows for more bombs to be dropped, then he can place more bombs up to the limit and no more due to fairness. All surfaces are relatively smooth inside of these arenas, so the power up for sliding bombs could have the bombermen put some sort of really slick compound on the bottom of their bombs in order to basically neglect friction and the bomb will stop come exploding. So if people decided to get together, call themselves bomberman and try to explode one another with bombs, it is physically plausible. It is certainly not an activity to test their sanity, but physics will not be working against them.

ELECTRICITY – I DEFY MY CITY AND PHYSICS

Some games have used electricity as a gimmick to make for a more enjoyable experience. Some games electricity plays a small role, others a great deal. This fascination with electricity had gotten to the point where Sucker Punch Productions decided to make an entire game called Infamous based around the manipulation of electricity. This game lets the player choose whether the city under the main character's control will become a place of pure corruption or a helping-everyone community. Cole is able to store and discharge electricity at his will by various methods such as: propulsion and floating, shock waves, absorption from street lights, electric grenades, shock people at large distances and many others. We will give Cole an exception and allow him to be able to store and discharge electricity at will, but we will still examine two areas of his abilities: being able to shock people at a distance and hover in the air by releasing electricity behind him.

Electric charge is like any form of energy. It wants to take the path of least resistance and be in its lowest state. Imagine there is a 500m lightning rod and a person in a field far apart from one another during a lightning storm. When the clouds collect a large amount of energy, it will be preferable not only for the person, but for the cloud to discharge through the lightning rod seeing as it is the shortest path with the least resistance for the electricity to travel. The path needs to be complete for electricity to travel. This is the reason that if a wire is only connected to one terminal of a power source, no charge will flow. So somehow Cole has been able to defy physics and control the path of his electric blasts.

There are several points throughout the game where the player is asked to 'shoot' a stream of electricity to attack someone falling off of a building or standing on a concrete arch. As stated before, the electricity needs a path to travel which will offer the least resistance and create a path to spread out its charge. If an enemy is jumping off of a building, he or she would be suspended and have not be an efficient path for the charge to travel. If Cole were to shoot electricity towards the airborne enemy it would more than likely find its way to a nearby electrical panel or pipe. His method of attack would only be effective if the enemy were the easiest source to travel through, which is not likely in a densely populated city with a lot of electrical and metal components.

We now will examine Cole's ability to propel himself and float for extended periods of time by emitting short streams of electricity from his hands. For the purpose of not completely discrediting this situation, we will assume it is possible to emit just short streams of electricity, rather than a complete path. With this exception, will Cole be able to float and glide using his electricity?

The only way Cole would be able to glide and float with this power is for there to be something in the air or near the ground which will repel him; opposites attract and likes repel. This means that with Cole's large amount of negative charge being emitted, there must be something which will have a large enough negative charge by him to repel and give him the boost necessary to float or glide. Due to the nature of the city, there are two types of charges surrounding him. Very large neutral charges in the air, ground, etc. which will not react at all. The other are the negative charges in different types of electrical equipment. It is very doubtful that any electrical equipment would have a very large excess negative charge though, seeing as that would damage the equipment. Even if there was, keep in mind that Cole's electricity would just be absorbed by the equipment and probably explode. The outlook does not look so good for Cole to be able to glide around the city. He is going to have to stick to walking.

Cole does have one ability which does actually follow physics quite decently though: his ability to drain generators of their power. If we return to the statement about electricity taking the path of least resistance, this applies to this situation as well. When Cole is drained of his stored charge, he is almost acting as a lightning rod would. Rather than electricity travelling through a transformer on a power line, it could make the jump to Cole's body to charge him up. He may be a more attractive force offering less resistance than moving through a transformer, so this ability including the exception of absorption powers is somewhat accurate. The distance in which he does it at times it debatable, but when relatively close to the source he is draining, this is quite feasible.

MAGNETISM – CLANK, WE HAVE A PROBLEM

Insomniac Games created a trilogy of games in the Ratchet and Clank series by the name of Ratchet and Clank

Future. Each one stars a furry, squirrel-like creature who is a Lombax named Ratchet and a little robot named Clank, plus a whole bunch of other hilarious characters. These games take place in futuristic settings where they have technologically advanced weapons and accessories. Amongst notable pieces of technology are the magnetic, rocket boots. These boots sport nifty little rockets in the soles to be used as a propulsion system, as well as strong magnets as the soles of the shoes themselves. Ratchet uses these boots quite often to speed from point A to B and to ride along metal rails. Are these feasible pieces of technology?

As was mentioned before, for anyone to jump, float, or change directions in air, there needs to be some sort of surface to repel from because air alone does not provide enough of a repulsive force. Insomniac games must have someone in their game engine department who knows something about physics. Ratchet is only able to hover slightly on the ground and propel himself forward when his boots are angled in the appropriate direction. These boots do not allow Ratchet to fly but allow for slight levitation, which is what would occur if someone were to make replicas of them. If the hovering mechanic works, lets see if the technology for grinding the rails is correct.

A lot of metals are not inherently magnetic. They need some external force to make them so. If this were not the case then people's spoons, knives and forks in their drawers would be absolute chaos. When a magnet is placed over a certain metal, say a nickel for example, they will attract to one another and we'll have a nickel stuck to a magnet. There are a lot of metals which are attracted to magnets, even though they are not inherently magnetic, the process of attraction is called induction.

Metals, like all atoms, have positive, negative and neutral charges inside them. The property that makes a lot of metals special is the ability to move the negative charge quite readily, which is why some metals make good conductors as well. This property is useful when it comes to explaining why metals are magnetic. Recall from before, we stated that opposites attract, so a magnet drawing ever closer to a nickel aligns the scattered poles inside the material to all be facing one direction. So when all of these poles align, there is an attraction which allows them to stick together. When the magnet is removed, the poles will return to a disarray due to it being a more stable form. They won't if the material has ferromagnetic properties, such as iron, where the poles may remain lined up to create a relatively permanent magnet.

Referring to Ratchet's boots. Using the means of propulsion and these magnetic forces, if he tilts his feet slightly to create a force to propel him forward along with a good amount of the boots on a piece of metal, it is very possible for him to be able to ride rails. There needs to be a good balance between the magnet soles and the rockets. They need to provide enough propulsion to allow Ratchet to go in the direction needed, yet enough magnetic force to keep him attached to the rail, but not stuck in one position. This could even be solved by having one side of the boots having a stronger magnet than the other side. But does this moving magnetic field on the rails create a new problem?

Recall that if we put a magnet on a metal, there could very well be a large accumulation of negative charge on the surface. Current (electricity), is the movement of these charges inside of a conducting material. When a magnet is moving by a surface, this induction will create a current in the metal. It is known that we shouldn't stick forks in the toaster or sockets due to avoid getting shocked. This may mean the end of the furry little Lombax which may leave the universe in great peril as Dr. Nefarious assumes control of it all!

Actually, Ratchet will be safe from being electro-cuted. After all, birds constantly stand on power lines which are a major source of electricity. They are safe be-cause they are not creating a path for the electricity to go to a lower state. If the birds were to place one foot on the power line and the other on metal poles there may be a problem due to them creating a path where the electricity can travel through them to get to the lower potential Earth. Ratchet is safe by the same principal. When he is moving along these rails, he never connects himself to a ground and when he passes a potential point to be electrocuted, he jumps over it. It looks like Insomniac Games did their homework. Insomniac Games may have done their home-work, but it does not mean that Professor Von Kripplespac did.

Conker's Bad Fur Bay stands out as one of the most memorable and unique games on the Nintendo 64. Swear-ing, booze, a scarecrow, fighting 'The Great Mighty Poo', what is not too love? Due to the plethora vulgar and ques-tionable situations, I cannot help but hold it dear to my heart. I especially love the wonderful multiplayer on the best multiplayer console available at the time. Since Profess-or Von Kripplespac gave it the go-ahead, we will examine how the anti-gravity chocolate works. If you do not like it, there is no sense getting upset over spilt milk (like a certain Panther King).

First off we will establish that Professor Von Kripplespac is a liar. If it were anti-gravity chocolate (assuming that actually means anything), then it would soar endlessly away from Earth and all other bodies which have a gravitational pull rather than floating slightly above the ground. But calling a spade a spade, "Induced-magnet-superconductor-plus-a-magnet-chocolate" does not exactly have the same ring to it. So what does it mean to be an IM-SPMC?

First we will establish more important points about magnets. Every magnet has a positive and negative side so to speak (or a north and a south pole). Two positive ends will not be naturally attracted to one another and with strong magnets you may not even be able to force them together by hand (likewise for two negatives). Whereas a positive and a negative end will naturally be drawn to one another. It is important to note that if you cut a magnet in half that you will not be left with a positive or a negative end, you will have two smaller magnets with opposing ends weaker in strength than the original. You can think of the inside of a magnet of having 'lines' inside all pointing in one direction to dictate which end is positive and which is negative. Now this is where things get a little tricky, but we will get through it as simply as possible with no potty humour at all. Even though jokes are easy when talking about a game with the Mighty Poo in it.

There are a number of ways you can make magnets float on one another: you can buy devices which do so or you can submerge a superconductor in a liquid nitrogen bath (or other cold conditions) and place a strong magnet over it to induce a current to create a dynamic magnetic field in the superconductor, as examples. For the sake of simplicity, we will be using the device which is electrically operated to allow a magnetic material to float above. The short explanation as to how this device works is that electricity flows through some type of conductor, which creates a dynamic magnetic field which will repel a strong magnet whilst simultaneously attracting it just enough to keep it floating in place. Electromagnetism is probably the most complicated and difficult area of physics to understand, hence the simplification. There are many places across the web to learn more, just don't start with Youtube comments (at least those ones where you lose faith in humanity after reading them). Now we need to apply this to our IMSPMC.

These chocolates are in rigidly prescribed locations, this must mean that those base units seen in the video are large, placed in the ground and painted to blend in. This also means that there is a large magnetic disc inside the respective chocolate in order for it to float in place. I guess when Conker devours these massive chocolates, he induces vomiting between areas in order to rid himself of them as to not stop the action and flow of the game. So do not overfeed him with magnets!

Although Conker and his friends do not obey common social conventions, at least they obey physics in some respects.

MAGNETIC FIELDS – SPACE LIGHTBULBS ATTACK

Thanks to 'Wiikid6' (obviously a pseudonym) for the suggestion for this particular section. We will take a look at electronic shields used in a shmup called Silpheed. A scmup being a contraction for shoot 'em up, which is a game where the player takes control of a ship and shoots things, pretty simple idea.

Silpheed and many other games have electronic shields which may or may not regenerate depending on the circumstance. Unfortunately in the case of Silpheed, the shield does not regenerate. I am sure many wish it did as it would make the game a whole lot easier, perhaps too easy. So how exactly do the shields get charged in the first place?

If you have watched any of Roo's "The Way Games Work", know anything about solar panels, or many other concepts, you may have already had an introduction as to what photoconductors are. The basic principle is that photons (light) will make contact with the surface of the photoconductor and be converted into a more useful electrical current. This electrical current can be used in various ways, such as powering the ship and more importantly in this case, powering the shields. As a side note, remember that energy can neither be created nor destroyed. This photoconductor process agrees with that principle because the light (one form of energy) is converted in the electrical current (another form of energy). Now that we know how the shield gets charged, let us try and figure out what the shield is composed of.

Rather than over-complicating this situation, we will use the electrical current in this form rather than converting it again. When electricity travels through a coil (a solenoid, which looks like a slinky made of wire), the winding path creates a magnetic field. We will assume this ship is wrapped in superconducting wire in order to create a protective magnetic field. As a precaution the inside of the ship is protected by an insulator so no electronics will be damaged. But if we have a large photoconducting surface and a seemingly infinite source of light like the sun, why does the shield in Silpheed not recharge?

My guess is that attacking ships are not actually shooting lasers like I initially assumed. Light moves at 300 000 000 metres per second (approximately 8 times around the Earth in one second), so the shots move way too slow to be lasers. Maybe the shots fired are actually light bulbs filled with tiny rocks in various shapes and sizes. When the shield stops the light bulbs (seeing as there is metal), the momentum of the rocks inside destroys the container and barrage the ship. Rocks (without metals inside) are not magnetic so they simply pass through the magnetic field no problem. The rock barrage explains why recharging is not possible because when photoconductors are damaged even slightly they lose all effectiveness due to being extremely fragile. Thus meaning this is a very stupid design which is not very effective. At least the ship will be protected if attacked by magnets and small pieces of metal.

ELECTROMAGNETIC RADIATION - STUPID EMR

Konami published Kojima Studio's incredible series known as Metal Gear Solid. Metal Gear Solid is a long running tactical stealth espionage series which not only defined the genre of stealth games, but has constantly stayed on top of innovation. Metal Gear Solid 4 (MGS4) is no exception to this rule as there is increased the intelligence of the AI, added sensitivity features for being detected such as the sound of a back cracking and smell. MGS4 in particular has been very thorough in terms of its physics mechanics including the heaviness of footsteps, smell and even jumping and climbing mechanics. There is only one scene in particular that its creators completely missed their mark. Everything up to the final scene of the game is very plausible to occur within a certain degree (SPOILER ALERT

COMING!!). The scene which we will analyze is when Snake is making his way to his final mission and is crawling through a giant microwave.

Electromagnetic (EM) radiation comes in many forms ranging anywhere from microwaves, visible light waves to gamma rays. When we refer to them as waves, they are in actuality, like waves we would find on a lake on a windy day. If the motion of the waves is constant (same height, same distance between them), then we can determine the wavelength of each wave. The wavelength is the distance between the same spot on two separate waves. As an example we can measure the length between the crests (tops) of two waves to determine the wavelength, which is typically measured in metres. Along with the wavelength we can determine how many times the wave cycles past a point in a given time, in other words, the frequency. Knowing both the frequency and the wavelength we can determine the speed of the wave in question using the universal wave equation: $v = f \lambda$; where the Greek letter nu (not v) is the speed, f is frequency and the Greek letter lambda indicates our wavelength.

Let's return to the EM spectrum. Light follows the same formula for our use. The speed of light is approximately $3.00 \times 10^{8} \, m/s$ (that is 300 million metres in one second, the distance to the sun and back in one second, remember from Silpheed probably moments ago) so using our wave equation, we can determine the frequency or wavelength of a given wave on the spectrum. We can view this equation as sort of a balancing act. If we have a wave with a large wavelength like a microwave (which can be a metre long), to balance this out we need to have a frequency lower than say a gamma ray, which has an extremely high frequency, but low wavelength. If we calculate that value for the metre long microwave we have the following:

$$3.00\,x\,10^8\,m/s = (1\text{m})\,f \rightarrow f = 3.0\,x\,10^7\,Hz$$

So if we recall what that represents from the case of the pendulum, that would mean this wave cycles 30 million times in one second.

Each type of radiation comes with its own inherent energy depending on where it is placed on the EM spectrum. Waves with a greater frequency, such as gamma or ultraviolet rays, have larger amounts of energy, vise versa for microwaves. Lower energy waves like the microwave still have a lot of energy, just not nearly as much as the others. Each type of wave also has penetrating and reacting abilities which are different as well. Take for example, ultraviolet rays can penetrate a person's skin and give them a pigment which is referred to as tanning (results vary by skin colour). Something like a microwave has different reactions to materials, such as people using them in their microwave ovens. Microwaves in a microwave oven can be particularly useful for people to warm up their food because the microwaves 'excite' water molecules. When the water is bombarded by these rays, the molecules receive energy which increases their speed and number of collisions. The increase in collisions causes the water to and heat up, which causes the water to expand and become a vapour. This is where Snake and his beautiful mustache get into trouble.

Snake in an epic struggle, slowly crawls along an extensive hallway which is completely surrounded by walls emitting microwaves (why he doesn't send his robot on this mission is a mystery). The human body is filled with a very large amount of water and when it heats up we have a change of state where liquid water turns into gaseous water. In a household microwave the steam will exit the food, letting water vapour move inside the microwave. Water inside Snake's body, however, has nowhere to go. Gases occupy more space than a liquid, meaning they will attempt to stretch out the surrounding container which happens to be Snake's body. His body would not expand like a gigantic balloon (like Kirby's) unfortunately, letting him waddle his way to victory. Even if his advanced stealth suit has an incredibly large amount of endurance and will not budge, this will not save him from his vital organs blowing up. Thus Snake wouldn't have lasted very long in that chamber and would have left the world to an impending doom.

One fact could actually save this scene from being incorrect. There is no disputing whether these are microwaves or not as the game frequently refers to them as such. But Snake may very well not be composed of as much water as a normal person seeing he is a bastardized clone from advanced technology. This means there is the possibility of Snake's body having some percentage of water which is being excited, but not enough to cause any internal damage. So his life lies in the hands of how 'human' he is. The more water he contains, the more invalid this scene is. For some strange reason the game never gives mention to his water composition.

We will now shift our focus to the visible part of the spectrum, with a wonderful survival-horror game called Ascension. Ascension was develop and published by Magnesium Ninja and can be found for free online, here:http://www.magnesiumninja.com/ . The stand out feature of this game is the use of the flashlight to examine surroundings. During the best segments of this game, aside from a strange glow being emitted faintly around the player and the flashlight, there is no way to see any of the surroundings, total darkness subsides. This allows for the mind of the player to build tension through imagination, rather than the developers trying to convince you what should terrify you. Anticipation and a wandering imagination are extremely effective ways of building tension. It is a brilliant tactic that makes for a wonderful game; plus it's free, so nothing is lost giving it a try!

Now onto the physics aspect, the reason why we are here. You know a flashlight works, but have you ever questioned how it works, or more importantly, why its beam fade away the further it goes? Let's first determine how a flashlight produces light.

As we probably know, light is produced from a light bulb. What happens is, when the switch is put into the ON position, electricity will begin travelling through the light bulb. This electricity will travel through what is called the filament, where it produces heat. Depending on the type of metal the filament is made of, a particular colour of light will be emitted; this process of heat emitting light is referred to as incandescence. Different metals will provide different colours, and will also require different amounts of current to produce the light. Now that we know how the saviour beam of light is produced, we need to know why to stupid thing doesn't light up the room and leaves us to face our worst enemy: our mind.

Light can travel through the vacuum of space, bend around corners, heat things up, why can't it even travel to the other side of the bloody room sometimes? Let's compare what is in space in contrast to our air to see where the difference lies. As we explored with Dead Space, sound is not able to travel in space because there are no particles in the air, but is able to do some on Earth because of the composition of our atmosphere. So the same thing that promotes the travel of sound from gruesome enemies, also hinders the use of a flashlight.

When light travels in our air, there are all sorts of little molecules for it to bounce off of, disperse and be absorbed by; examples include water (where rainbows are born), carbon monoxide and nitrogen. So unfortunately for our hero in Ascension (I don't even want to ruin the name, go play it, it's free!), the developers knew something about physics and didn't let the flashlight project a beam of light indefinitely.

SECTION III – MODERN PHYSICS
QUANTUM TUNNELLING – I CHOOSE YOU
QUANTACHU!

Quantum mechanics and classical mechanics do not get along well with one another. There are many concepts which classical mechanics say are impossible and quantum mechanics goes around classical's back and tells everyone how wrong it is. It is like the know-it-all who corrects a person when the say one incorrect word in a movie quote. Although the know-it-all is correct, it is not very nice to continue to point out faults. One of the issues that classical says is impossible, but quantum mechanics does is quantum tunnelling (since it is not called classical tunnelling, it is probably pretty obvious). Prior to examining that concept, we will examine what is so different about the two types of mechanics.

Classical mechanics is intended to explain phenomena happening on a larger, macro scale. It helped us explain why Wander could climb a colossi or a soldier could jump a gorge on a snowmobile, pretty well everything up to this point in many respects. Quantum mechanics on the other hand, deals with the micro scale. It helps explain how individual particles travel and react with one another. Quantum theory states that all matter acts as a wave as well, so even people walking around have a wavelength. In contrast, classical mechanics if we recall, only accepted light as dual matter due to it refracting and acting like a sound wave, but also reacting like chemicals found in a lab. Classical mechanics would never predict or explain why a person would bend around a corner like a sound wave simply because it does not happen. This duality that quantum mechanics supports explains why quantum tunnelling can occur.

Quantum tunnelling is not just some theory which has come about with no examples to prove its existence, like many people seem to believe. An example of quantum tunnelling is radioactive decay. When an unstable uranium needs to become more stable, it will have certain particles tunnel out pending they have enough energy to do so. Both quantum mechanics and classical mechanics have different analogies which can explain how this works.

There is a common example of how tunnelling works using classical mechanics (just to prove classical and quantum mechanics can play nice together). Tunnelling is similar to someone attempting to roll a ball up a hill. If there is a smaller hill, less energy has to be used to get the ball to roll up, the larger the hill, the more energy that needs to be put in (keeping in mind that energy correlates to force). If there is a sufficient amount of energy and the orientation is correct, the ball may roll up the hill and down the other side. Orientation is important as it would not matter how much energy is put into rolling the ball if it was pushed away from the hill. This is similar to how, say an electron, tunnels through a barrier.

Quantum mechanics takes a similar approach to the classical one, but rather than a ball and a hill, it could be an electron going through a 'barrier'. If an electron is approaching a barrier it can penetrate, if given sufficient energy and the wave movement of the particle is correctly orientated, tunnelling will occur. There are ways of using equations, wave functions and advanced calculus to solve for the probability of this happening which can be found on internet discussions and in quantum mechanics text books. We will just focus on the concepts involved with this instead of the mathematics.

We can compare this electron tunnelling through a barrier to a rubber ball getting thrown through a hole in the wall the same diametre of the ball. If a person were to continue to throw the ball against the wall hard enough with the correct orientation, he or she could eventually throw the ball through the hole in the wall so the ball could be free from abuse on the other side. The odds of this happening are very small, but it is certainly possible to. If instead of one, we have thousands of electrons attempting to tunnel, the likelihood of tunnelling to occur (depending on the circumstances) is much greater due to the enormous amount of times an electron is able to attempt tunnelling in even just a second. In one second, an electron can attempt tunnelling well over trillions of times. Now we need to move onto why we even bothered with this in the first place.

Developed by Game Freak and published by Nintendo and The Pokemon Company, Pokemon (short for Pocket Monsters) Soul Silver and Heart Gold are part of the most successful RPG series of all time. This analysis could be done on any of the games in the main Pokemon series as the same theories apply, but seeing as Soul Silver is my favourite in the franchise thus far, we will use it instead. The basic premise is for the player to capture pokemon and train them to be the strongest pokemon they can be. The test of the true trainer is to go to gyms located all through out the region and duel with the gym leader and friends to see whos pokemon are the best. The ultimate test of proper training is for the player to beat the best trainers known as the elite four and the champion after beating all of the gym leaders. The more advanced player will try to capture at least one of every type of pokemon to show they can amass an impressive collection.

The player has the potential to capture a pokemen by using a pokeball. To capture the pokemon, it is shrunk and put into a ball which can conveniently fit in ones pocket of latch onto his or her belt. The player can continue to let his or her collection of pokeballs containing pokemon grow until they have hundreds deaming them the ultimate collector (ignore any ethical considerations, it's for the best). This capturing processes involves two concepts, one ludicrous, the other just highly unlikely (assuming the Pokemon universe is plausible).

The first part of the capturing process is shrinking whatever pokemon it is, whether 12 or 2 metres tall, to the size of the inside of the pokeball. This pokeball can then be placed securely in a pocket, reinforcing the pocket monster short form. The concept of shrinking while maintaining the same form and composition is absolutely ludicrous. If something becomes compressed, it must change its composition. If a gas is compressed, under enough pressure it can become a liquid, with even more pressure it can become a solid. There are the same amount of molecules and same types, but their arrangement definitely changes. If we try and compress a 12 metre tall pokemon into a size of a persons pocket, we are taking a solid and making it an even more tightly bound solid. There is no way that the structure of this pokemon would remain intact.

Let us assume that this is not a compression, but like taking a picture and resizing it so it will look the same, but be smaller. If we are to resize a picture in an image editing program and make it smaller, it typically works in the following way. Let us assume we have a square divided into four equal squares and each of these squares is the same colour. If we remove three of those squares, we are left with one which was identical to the three that were lost. This is sort of how resizing a picture works. We take a pixel (our bigger square) and eliminate parts of that pixel to be left with a smaller amount (our single small square). If we open a photo and zoom in really close, we can actually see these pixels. When we resize a picture to become smaller, the image will resemble the original picture in some form, but not perfectly. The smaller the picture gets, the more data is lost. This is why taking a small picture and trying to create a larger image does not work, because the program is trying to compensate for lost data. Some programs have incredible features built in to do this quite well, but it still will never be perfect.

So this could be the other way which a pokemon could fit in a pokeball. It could basically act like a really good picture editing software and shrink and grow, eliminating and creating data. This could be the way it would work if pokemon were photos and not living beings though. Seeing as they are living creatures, this would mean that they would basically shed massive parts of their body and genetic makeup to be a scale look alike of their former self, but not be the exact same. When they are released from their pokeball home, they have to rapidly reproduce all of their lost body components to become themselves again. This does not make any sense whatsoever. However, so we can examine how the quantum tunnelling works.

Whenever a pokemon returns to their pokeball, something which resembles a laser is fired at the pokemon, the pokemon shrinks and they go inside the ball without it ever opening. We can see clearly ball itself is a solid sphere with no holes in it, so we are going to assume that the pokemon does the same things our electron did with its barrier and it tunnels. It would be very surprising if this actually occured though, let us see why.

Let us return to the example of the person throwing the ball against the wall trying to get it through the hole. Imagine instead we had billions of these balls and had to throw them through billions of holes. We can think of a pokemon being made up of billions of those balls seeing as all matter, all living beings are composed in a similar way on an incredibly small scale. We can see that all individual particles are close to one another bound by forces, but not touching one another. So if we want a pokemon to tunnel to the inside of the pokeball, we need to have all of those billions of balls go through the wall at the exact same time. This would be assuming a two dimensional form of the pokemon, but seeing as it is three dimensional we would need to have waves of these balls to go through at the exact same time.

The more energy we give to these balls, the more probability they have to go through the wall. So if the laser beam that returns to pokemon accelerates them to the point where they are basically going as fast as light, they would have more of a chance of tunnelling. But the likelihood of this occurring is incredibly, incredibly, INCREDIBLY small. A person would have just as much luck trying to tunnel through a concrete wall without harming the wall. Although statistically speaking it could happen, our sun will probably burn out and the Earth will seize to exist before that could ever occur. Tunnelling does occur on a smaller scale all of the time, but on a macro scale it is just not plausible, at least not with the technology we have available at the moment. I broke my Pokeball and cannot test this; but statistically it does seem possible at a glance for a Pokemon to tunnel through the wall of a Pokeball. The idea and the technology are well beyond our grasp at this time, but maybe one day it will be possible. For the time being I have to continue my quest to be the very best, the best there ever was... or something to that effect.

TIME TRAVEL – FROG'S THEME, INFINITELY GLORIOUS

Ask anyone who knows anything about video games and they'll tell you Chrono Trigger is just solid. If you like RPGs, odds are you like (or will like) Chrono Trigger. That is unless of course you are one of those people who hates something just because too many people like it, in which case, there's no hope for you. Chrono Trigger contains all the usual suspects of great game elements: solid music, atmosphere, characters, fighting system; everything about this game is polished and refined. Reviews of this game range from proper tidy to bloody fantastic. This view of the game is not even impacted by nostalgia as I played this game for the first time in the late 2000s after all of the initial hype had long since died down, so there. Chrono Trigger stood the test of time (it's a pun!), but can it hold up to some science?!

We are going to dive right in and talk about time travelling. Do contemporary theories support the concept of time travel? To answer this we need to understand a little bit about the dimensions in our universe.

We can think of time as one dimension among several. Our universe is composed of four dimensions total: up, down (1, typically y), left, right (2, typically z), forward, backward (3, typically x), everything in between and time

(4, typically t). This is unless you believe in that silly string theory which can have dozens, such as a dimension which is saddle shaped; we will stick to four. There is an equation using the Lorentz transformation which relates all of these terms together (Beiser, 2003):

$$(\Delta s)^2 = -(\Delta x)^2 - (\Delta y)^2 - (\Delta z)^2 + (c\Delta t)^2$$

Where c is the speed of light and s an arbitrary inertial frame. So basically, we are in the middle of two 'dimensional cones' indicating we exist in the present time, where the cones meet, the cones simultaneously representing the past and present. To simplify the argument (as simple as we can get it), we will reduce this to just the x-dimension. However, the same assumptions apply if the other dimensions were included. So we are now left with the following:

$$(\Delta s)^2 = (c\Delta t)^2 - (\Delta x)^2$$

We will define s a little bit more: s is an inertial frame. This frame has to deal with a particular position in time. On Earth, many assume time to progress in a linear fashion at a constant rate. However, there are many conditions which will affect the way time progresses with reference to another frames; travelling towards the speed of light, time travels at a much different pace, thus meaning time is not linear, showing one of the conditions that effects time. Based on these inertial frames, if someone were to travel faster, or even close to, the speed of light, they would progress differently through time with respect to someone walking to work that morning. Some believe these ideas to be total bunkum, but others recognize how some use them regularly in their day to day life for calculations, like GPS systems (the more you know!). What does this mean for Chrono Trigger?

The vessel which the crew in Chrono Trigger uses to

travel through time will travel faster and slightly slower than the speed of light in order to progress backwards and forwards through time respectively; I would show you mine, but it's in the shop. However, this introduces a new problem. Even if this could represent one way that the ship could travel through time, the ship's crew would be subjected to time progression and age to the point that they would die or be unborn. This means that Frog might turn into a lil' ol' tadpole! If we pretend that the ship somehow isolates time progression, then this theory may work, but there could be another explanation. Actually I lied, it is not how the theory works, so I was a little misleading to entertain the idea and talk about Chrono Trigger a little bit more; the game deserves so much attention.

Relativity dictates that time will progress different with respect to another object based on certain conditions. Returning to the example of someone travelling at the speed of light, the other walking, one is not time travelling in the sense that they basically teleport to a moment in time, time is just progressing different relative to the other individual. This means that the time travelling vessel would not be able to travel to these isolated moments in time in order to save the world. However, there is another theory which may possibly solve our dilemma of conjuring an explanation.

Another idea, which falls out of some versions of that silly string theory (which isn't really silly, it's actually quite rigorous and countless brilliant people have worked on it) previously mentioned, supports the concept of there being an infinite amount of infinitely extending universes. This means that, although incredibly unlikely but statistically possible, there are planets very similar to the one which

Chrono Trigger takes place on. This means that rather than travelling through time, each time the crew goes in the ship, we are viewing the events of a different but very similar planet in another universe. Rather than events in the past creating paradoxes, we have an infinite amount of planets with events progressing as if the events in another universe happened.

So while the crew is desperately fighting a foe, hanging on by a thread for the fate of their world, another universe has the crew eating muffins and joking around with one another. If this applied to the real world, maybe I am writing this book in this universe and in another, there is a replica of me proclaiming that over-complicating and over-analyzing video games as a hobby is total hogwash. I hope not, because that would be rude. I am looking at you angry people leaving me nasty comments and e-mails, formally known as trolls. We use the most formal terminology here, anything else is flapdoodle.

DIMENSIONS – 3D DOT TRUTH TELLERS

3D Dot Game Heroes, developed by Silicon Studio and published by Atlus, is one of those games that has a tremendous amount of charm and personality due to it's use of voxels (basically looks like cubes), which pay tribute to retro games. Other aspects that standout are the plethora of fun, customizable weapons and characters, heart-warming story, hilarious tribute jokes, brilliantly composed soundtrack, outstanding level design and variety in gameplay. Not enough good things can be said about this game; it is a perfect example of how something written to pay tribute to numerous titles, can still provide enough content to be greatly appreciated by someone unknowing of the references. What is most important to us with 3D Dot Game Heroes, is that while paying tribute to retro games, it exposes their lies as well.

Many of the titles 3D Dot Game Heroes references, the retro ones at least, are considered to be 2D gaming, meaning two dimensions. Many retro titles and recent retro tributes seem to be classified as 2D games as well. While these older titles, like Kirby for example, lack the voxels or other ways of representing 3 dimensions for the players to explore, limiting them to 2 dimensions of exploration; this does not mean they are 2D games by any means.

Recall what was said about dimensions in our universe, we have four of them, often represented by the variables x, y, z and t. Given that most people reference dimensions only in terms of physical movements, we will forgive the exclusion of time being a dimension and assume we are only referring to physical dimensions we can manipulate. Games that seem to be considered 2D often just include the x (forward & backward) and y (up & down) dimensions for the player to be able to move in, but they pretend the z (left & right) does not exist; this is certainly incorrect to pretend.

Although the movement of the player is restricted to the two dimensions, the games should still be considered 3D. If they were not a 3D game, they would be pictures, something completely immobile. Without any depth (our lovely z axis), the player would be rendered helpless as the character he orshe is playing as would be a part of both the back and foreground. Even if we hypothetically allow for this flat world to exist and the player could move the character somehow, try to think about this from the character's perspective. Returning to the example of Kirby, what would the little puff-ball see in this flat world? An infinitesimally thin line, with nothing being distinguishable. Thus, Kirby and all other concerned characters, would have to live the equivalent life of a blind person, where all experiences would not necessarily be worse, but experienced in new ways. But given the way all of the characters in these games act, they are not blind, nor do they face these differences. So gaming has never really been in two dimensions in spite of what some may believe. Super Paper Mario gives players a behind the scenes look at what '2D life' really is.

Developed by Intelligent Systems and published by Nintendo, Super Paper Mario is a game which contains many of the '2D' game elements, where the player goes through platforming sections with different characters, mostly taking advantage of the x and y dimensions (the paper part kind of gives that away). So Mario and friends appear to follow the life of the perceived 2D character, but when Mario hits a dead end, he is able to take advantage of the z dimension and switch where the camera is focused, bringing depth into the equation. It is nice to know that there are '2D' games that acknowledge what allows them to not be pictures.

TELEPORTATION – IT'S A LIE

"Good news. I figured out what to do with all the money I save recycling your one roomful of air. When you die, I'm going to laminate your skeleton and pose you in the lobby. That way future generations can learn from you how not to have your unfortunate bone structure" (Credit to Test Chamber 12 in Portal 2). Actually, that isn't good news at all. Since you and your skeletal structure are quite acceptable, let us learn a little bit about this wonderful puzzle game released by Valve.

Portal is a series developed and published by Valve (unless on consoles, then it is published by Electronic Arts). These games have been an enormous success due to their unique puzzle solving mechanics and ridiculous sense of humour. There have been many jokes in the series which have become fads commonly referenced on the internet (the title of this section might be one) as well as many moments which have players laughing ridiculously hard in the confinement of their home. It is a rather sophisticated humour which is shared by very few games separating it from many others. The other factor that makes these games so distinguishable from others is that the puzzles involving problem solving with portals.

The player is able to use a portal gun which has two different portals it can shoot as a means of passing an object through one and letting it fall out the other using none other than teleportation. If there is a button way up on an unreachable ledge, we can fire one portal on the roof, another on the wall to drop a companion cube or even our fleshy body onto that button. If there is a turret on the floor that is bothering us, we can put a portal under it and have the other on a far away wall above a pit to send it flying to its doom. There are some surfaces which do not allow portals, but there are many places we can use portals to have countless hours of fun.

Teleporting: what can this possibly have to do with physics? There are theories and rumors of teleporting actually being a feasible phenomena. Yes, teleportation is real, you heard me right. Unlike Glados, I would never try to steer you wrong... unless you crossed me, in which case I will take you down when I invent a combustible lemon. Scientists have teleported particles over short distances in con-

trolled experiments. Before we proceed, we need to understand a little more about what teleportation actually is.

Let us pretend there is a troll on the internet who is
getting ready to send a raging, irrelevant e-mail to a video
creator. I do not think anyone would actually do that, it
would be rather cruel. Said troll types up ridiculous comments, presses send and the creator gets them and completely ignores them as they do not matter in any respect
what-so-ever. How did the comments get there?

The troll typed up the message from a completely
different location and did not physically show the creator
the screen and press their face against it in an assertive and
quite frankly, rude manner. The trollolo somehow managed
to squeeze the ridiculous comments into some cords, send
it through more cords, some servers, possibly some other
cords, wireless signals, then probably some more cords, until reaching the creator in a ridiculously short time frame.
The initial message was translated into code understood by
a series of media in which it travelled, changed forms and
eventually reconstructed in the a seemingly identical form
to the original, assuming we do not have trickery or a
purple monkey dishwasher situation on our hands. The
communication between these computers is based on the
deconstruction and reconstruction of vital pieces of information. Teleportation work in a similar way, but with one major difference.

When dealing with the concept of teleportation, we
are following that same principle of communication
between two different locations, but this time without having any sort of physical movement between location A and
location B. Non-physical is good, last time I got physical I
wound up sending a "turkey soaring majestically like an

eagle piloting a blimp" (Credit to Test Chamber 9 in Portal 2); it was a rather awkward situation to try to explain to explain why it happened to my family during the holiday season. This principal of teleportation indicates that there can be no travelling between the initial and final location, just an instant transmission of the matter in question. Point A is supposed to deconstruct and determine the arrangement of the matter, then communicate that information about the initial matter, then reconstruct at point B. The key word being supposed to; when I was doing a test run of teleportation for this section, I wound up losing my left shoe, half the right side of my pants, the collar of my shirt and the hair from my knuckles, bloody contraption... I still have no idea where any of them are either.

Teleportation has already been done through experiment with particles. Let us examine the basic idea of this without the rigorous mathematics that can be involved and teleport ourselves an electron.

Let's assume the only two things we need to know about an electron, its charge and orientation. At point A, this information is known and it is communicated to point B. Point B learns about an electron which has a particular charge and orientation. So the electron at Point A deconstructs while simultaneously reconstructing at Point B in order to satisfy the conservation of energy. By quantum mechanics standards, this means that particles, people and companion cubes are all just forms of information. Yes, you are information. Glorious, wonderful information. Even you potential troll readying the e-mail, I love you too. Rather than worrying about one little particle, we have trillions and trillions, actually considerably more, to worry about with Portal's turrets, cubes and people.

Assume we have one orange portal on the roof and the other blue portal in front of our heroine as she clenches a cube for her dear life; if she decides to say farewell to her dear companion and toss it through the portal, this follows the teleportation and communication concept. When the companion cube hits the blue portal in front of our troubled character, it becomes transmitted to the orange portal on the roof. This happens instantly as the blue portal communicates to the orange portal the characteristics and make-up of the cube as it passes through. This is all happening instantly to allow for a fluid teleportation and conservation of energy. If going by some more recent theories which seem to be surfacing, Portal could very well be a potential real simulation in the future rather than just a video game. Maybe in the future, we can torture poor souls with these theories. Try and have them adapt and create some combustible lemons. There is another game where hilarity ensues and teleportation is completely acceptable.

The generally hated Eat Lead: The Return of Matt Hazard, which is a shame because I love it and think it's hilarious, has teleportation which is acceptable within the game by our current understanding of how to teleport objects. Why is it acceptable in Matt Hazard, but not Portal?

To answer this we do not have to examine very much as Matt Hazard is a video game character, thus he is simply code much like any data contained on a disc or computer. Yes, Portal is a video game as well, but Matt Hazard is in a different sense. Eat Lead follows the story of Matt Hazard who is a self-aware video game character, it is the entire premise of his character. Thus, Matt Hazard can be treated basically like a file that a person moves from location to location on a computer. He can even be copied and tinkered around with should the person please. Now only if we could take the principles presented in Matt Hazard and apply them to some combustible lemons, the world would be a very different place.

WAVEFUNCTION PARADOX – YOU MURDERER

Why not close this section off with some controversy and what some people consider to be the most emotional moment in gaming ever. Thus meaning, if you have not played Final Fantasy VII, massive spoiler alert, skip this section. If you have played it, you probably know what we are covering, the death of Aeris in this classic RPG which is held in high regard like Chrono Trigger.

There have been many hypotheses regarding the revival or preventing the death of Aeris. Do some random stuff with a ghost, fulfill ridiculous conditions to prevent the death as a whole, none of them work. What I am here to tell is that Schrödinger's cat says you killed Aeris. Not Sephiroth. You killed her by going to see her.

Let's lay down some groundwork about what the paradox Schrödinger suggests with his cat thought experiment (Griffiths, 2005):

A cat placed in a steel chamber, together with the following hellish contraption... In a Geiger counter there is a tiny amount of radioactive substance, so tiny that maybe within an hour one of the atoms decays, but equally probable none of them decays. If one decays then the counter triggers and via a relay activates a little hammer which breaks a container of cyanide. If one has left the entire system for an hour, then one would say the cat is living if no atom has decayed. The first decay would have poisoned it. The wave function of the entire system would express this by containing equal

parts of the living and dead cat. (p. 430)
If one were to open the chamber to determine whether the cat is alive or not, this curiosity is what would determine the result of the experiment. Opening the chamber may cause the poison to escape and keep the cat alive, or kill the cat due to agitation of the unstable, radioactive substance. Let's try and clarify this with two other metaphors.

Let's say we have a box encase a vacuum, except one solitary particle. We do not know what this particle is, all we know is that it is inside the box. We know not of the charge, mass, speed, momentum, position, orientation, spin, nothing. The only way we are able to determine anything about this particle is to measure it in some capacity. But what happens when we put a measuring device in the box? We have now interfered with the particle. We may have changed its position due to some attraction, maybe its mass, there are so many factors that can change due to the curiosity. The error introduced by the measuring device could be accounted for with another device that corrects the error, but then what accounts for the error introduced by the error-correcting device? It is a perpetual cycle with no end.

So without being able to measure the particle, we have to make certain assumptions that would be impossible to determine. Take for example, we can claim that the particle is one of three charges: positive, neutral and negative. Which of the three, we cannot say for sure, but we can consider it a superposition of all three characteristics. We are not saying it is all three, it could just be one of the three. The act of measuring this to know for certain could change the outcome; or kill the cat so to speak. Let's apply these ideas so something a person could actually see day to day and stretch the idea pretty far, mostly regarding the idea of interference will change events.

If someone were to rudely trip you while you were walking down the street, that would interfere with the progress you would otherwise make while walking. You may have tripped outside of the person doing it to you, but we can say for certain that the person tripping you changed the situation entirely. Or if you were to go into a mall and yell "Final Fantasy VII is just ok, Final Fantasy IV, VI and IX are much better", even people who are not aware of what Final Fantasy is would certainly be affected by this scenario. It may not be a significant influence, but it is an influence nonetheless. So what does any of this have to do with passing the blame to the player for killing Aeris?

We can think of the area where Sephiroth and Aeris are to be an isolated location, much like the steel chamber containing a cat. In this location, Aeris has a wave function where she is equal parts dead and alive. If we are following the ideas we previously elaborated on, if we want to know whether she is dead or alive, we have to interfere with this isolation. As those who played this game can attest to, the interference definitely made Aeris a lot more stabbed in the back and a lot less alive. So if you want Aeris to live forever, do not progress in the game and she'll be just fine. Seriously, she will never die, that is the only undisputable way to ensure she lives. But should you be curious and have the slightest inclination to continue, just remember the old adage: "curiosity killed the cat".

SECTION IV – EXTRAS AND CLOSING NOTES
CONTORTION AND MORE – METROID
EXTRAVAGANZA

Remember way back when we said we were going to cover more Metroid; we are now going to return to the Metroid series with the original trilogy of games. Since I do not feel that praises for this series have been sung enough, we will continue to do so. Samus stands out to me as one of the best video game characters ever because she assumes a non-sexualized, powerful, independent role. She exhibits an overwhelming amount of strength and plethora of survival skills, both of which she uses to overcome seemingly insurmountable challenges and obstacles (remember we are in the mindset Other M does not exist, just considering the brilliant trilogy). To compliment this well-crafted heroine, the games themselves are outstanding experiences.

Players get to experience an emotionally gripping story of survival, persistence and determination, immersed in diverse environments where danger could be lurking at any moment. Gruesome and numerous bosses and enemies continuously challenge players to adapt fighting tactics in order to succeed. Combined with precise controls, dynamic and well-balanced soundtrack, carefully composed maps used for intuitive exploration and very compelling story without extensive dialogue; this is an incredible action-platforming series which is virtually unparalleled. Now that I have confessed my undying love for this series, let us see if Samus' morph ball is physically acceptable.

Free from the constraints of her bulky armour, being able to roll up in a little ball is certainly not unreasonable. Contortionists are capable of doing some pretty unbelievable things, pushing their bodies to inconceivable limits after a lot of training. If you want to learn more, look online for a Game Theory episode which talks about Samus and her precious morph ball (thanks MatPat for letting me offer extensions). Confined to her armour it is possible for Samus to contort the way she does as her armour is rather compartmentalized and may form around her so she can curl up. As a supplement to that fantastic video, we will explain why she is constantly rolling, never dizzy, able to see where she is going and able to bounce while curled up in the morph ball.

What we will propose is that Samus becomes gyroscopically stable confined in the centre of her armour. This means that no matter how much the morph ball spins she will remain stable in the centre to ensure she does not puke inside her armour (an act which would probably bring forth great discomfort). But how does she spin if she is isolated in the centre?

The outer 'shell' of her gyroscopic armour has a track which is constantly spinning and will vary how far it is protracted depending on whether she wants to spin in place or make some horizontal progress. There is a good reason for the track being at varying elevations. Slipping in place occurs when there is some friction, but not enough to move forward much like a slipping wheel on ice, this is when the track is retracted. When the track is protracted there is enough of a 'gripping' force on the ground to allow for movement to occur. In order for Samus to maintain control over her forward movement her armour is equip with external cameras which feed into a display inside. Finally we need to examine the ability to bounce.

Lucky for us the the game lets us know how it works seeing as the morph ball gets an upgrade called the 'spring ball'. We have made considerable and unpredictable progress when it comes to the progression of technology. If anyone told a child playing an Atari 2600 they would be able to use augmented reality cards on a portable console in the near future, that kid would probably be in total disbelief or completely lose it with excitement. The possibility of a spring being able to launch someone in heavy armour does not seem like that much of a leap in logic. Now time to finally address the elephant in the room.

We examined how this suit works with respect to the morph ball, but pretty well ignored the design and engineering aspects. There are a couple of reasons for this.

Number 1 – I really wanted an excuse to talk about one of my favourite games/characters of all time.

Number 2 – As physicists (I am dragging you down with me), we come up with crazy designs and ideas and get other very talented individuals to make our work practical and feasible. Being a physicist does have its perks (it's a secret to everyone though).

Let's continue our coverage of the morph ball by checking out the spider ball introduced in Metroid II. With it being a Game Boy game, this part of the trilogy seems to be under-appreciated and not talked about as often, which is a shame. This game is quite aptly named as a member of the Metroid series as Samus' primary mission in this game is to eradicate all Metroids on their home plant. Rather than just fighting the green and red, jellyfish-like Metroids typically associated with this series, there are a number of different and difficult Metroids to destroy, like the terrifying Omega Metroid. Rather than focus on why those things terrified me as a child, we will move onto our analysis.

We are going to cover the possibility of one of the

game's most important upgrades, the spider ball. We will assume the general design of the morph ball we proposed holds true and offer two different designs for the spider ball to exist. It is unfortunate we are not looking at the spider ball from the Metroid Prime series, as it can only be used on designated magnetic tracks, life would be easy then. But since life, much like this game, is not easy, we will have to put our thinking caps on to progress with our analysis.

Our first design involves a series of many electromagnets inside of a track, parallel to the one which allows Samus to move. An electromagnet acts similar to the magnets discussed with our "anti-gravity" chocolate, but only do so when they are being supplied with electricity. What this means is that magnetic attraction will stop as soon as deactivation is requested by the player. We will assume there is an insulating track to protect the electronics within the suit. Why choose magnets?

A part of this is answered easily, as there are a number of spots which contain metal structures, which we will assume to be ferromagnetic (attracted to magnets). However, there are a lot of spots which are not metal, but are rocks, minerals and such. We will assume that these cave walls are full of many unmined, ferromagnetic metals. These metals in conjunction with extremely powerful electromagnets could allow for Samus to adhere to walls, thus letting the spider ball be physical acceptable. But what if the walls are not full of these metals? Our question leads us to our second design.

Let's explore a literal interpretation of a spider ball by using spider silk. Spider silk has some amazing properties. Not only is it extremely durable, it has tremendous adhesive properties as well. Spider silk has what equates to a quick-drying super-glue on it, hence why spiderwebs can be annoying to take down and are capable of spanning over such great lengths to awkward locations. So maybe in place of the electromagnet track, there is a trick of spider silk which can be extended and depressed much like the spinning track discussed before; this will allow for quick deactivation much like the previous design.

Now that we stuck to our commitment, we can abandon our thinking caps for the time being, we deserve it. It is too bad we did not talk about spiders while we had the chance. Spiders love reading on websites, the world wide web concept is wonderful to them. Too bad we did not stick with it, but I do not know much about spiders and would not want to spin you a web of lies. No more terrible puns, those were just awful. Let's just move along to the final stretch of talking about the morph ball, as well as talking about some weapons and expansions while we're are it, just because we can.

While examining these weapons, we are going to have one exception guiding them all: unlimited ammo is going to be acceptable for the time being. If we were able to disprove the conservation of energy and create resources from nothing, we would not be here right now; we would be releasing articles that would shatter everything accepted in physics to date. Well, time to talk about many weapons from the original trilogy.

Bombs. How do bombs let Samus bounce up and down unharmed? Well, the bombs send out shock waves, which are dense packets of air. Being densely packed air would allow for there to be a substantial enough force to move Samus in her suit, but not cause damage to it. She is able to launch upwards in some fashion because the game only shows the player half of the situation. Bombs appear to be planted right beside Samus, but they are actually planted on both sides of her. With planting on both sides, side-swaying is rendered impossible and the forces are balanced out, thus leaving no where to go but up.

High Jump. This works very similar in principle to that of the spring ball, a series of extremely tense springs are at Samus' feet, aiding her to jump to new heights. Nice and simple.

Missile. These work like a RPGs (rocket propelled grenade), those exist already, no commenting really needed as to whether they work or not, but lets quickly observe how they work. Returning to the principle of every action having a reaction, what happens is there is an explosive reaction inside the arm cannon with some venting. The reaction causes a strong force that pushes out the back-end of the missile, this in turn causes the reaction to the action forcing the missile to go forward. Due to extreme force and short distance that the missile goes, it does not have time to visibly plummet downwards. As soon as the missile hits an enemy, it triggers an explosion from the explosive inside the casing.

Super Missile. The super missile is equivalent in power to five regular missiles. To use an analogy for all house repairs growing up, the old saying "if it doesn't work, get a bigger hammer" applies. The missile works the same, but with more explosive power.

Speed Booster. We return to the principles brought forth with Crysis regarding the exoskeleton being a source of enhancement to the human body. Rather than invisibility and mechanical advantage, we are going to hypothesize a design which will allow Samus to run at the tremendous speeds she does. Our proposed design is going to have a series of hydraulics at each of the joints used while running. This means that Samus has an aid to increase her speed.

The glowing on the suit comes from an energy conversion. When reactions take place, sometimes light, heat or other energies are emitted. Since the hydraulics are placed internally, heat will be gathered inside and will need a place to expel. The suit converts the extreme heat into the wonderful glow emitted.

Varia Suit. Insulation.

Gravity Suit. On top of doing something as wonderful as reducing damage taken, this suit offers the ability to jump in water as if jumping in air. To do this, there is probably a series of propulsion systems in chambers of the suit which can pivot based on the desired direction of travel. This includes being able to fall back down as if water is not around. Seeing as this suit is in Super Metroid and we do not have a very closeup look of it, we can't rule it out.

Ice Beam. This is where things get a little bit complicated. Unfortunately the explanation is not as simple as shooting ice blocks as those would hit the target, causing a minor inconvenience as it hurts to get thumped by ice. It might even cause a slight concussion or death if it hits a meningeal artery equivalent (probably not considered a minor inconvenience). But this certainly would not freeze an enemy. To get an explanation, we have to create some sort of capsule containing a chemical compound and two other capsules inside the larger one. The two smaller capsules are also filled, one with a chemical compound of sorts, the other a blue light. The blue light is a little more obvious, but the chemical compounds, not so much.

Remember endothermic reactions? If a reaction occurs and heat is absorbed and converted to other forms of energy, thus the final product will feel cold, it is considered to be an endothermic reaction. This is what the capsules are for.

Substances are often stable when isolated, mixing them together will create a reaction almost immediately sometimes; like a vinegar and baking soda rocket. When the large capsule hits an enemy, it causes the other capsules to break, which is when the light breaks and the substances start the endothermic reaction. With the help of a catalyst, the reaction will occur quickly enough to create what seems to be an instantaneous freezing.

Wave. X-rays, gamma rays, light, microwaves; each of these are waves as we explored in the case of Metal Gear Solid. Given Samus' height and the length of the wave pattern relative to that, we can assume that the wave gun resembles a microwave. Maybe those who constructed the gun found a way to concentrate some form of energy which travels the pattern of a microwave, but at a much slower speed. The wave beam is able to pass through walls and such, so it follows similar ideas and limitations to that of the Pokemon going through the walls of a pokeball.

Screw Attack. Uhhh....

DNA – KUNG FU JET SWORD BLOB

Since Kirby is such a wonderful blob to analyze when referring to his physique and abilities, we will take another look at him. This time we will examine his ability to clone the abilities of the enemies he swallows and decides to digest. This is actually a lot more plausible than it would seem at first glance. Let us use biology to explore this concept a little bit deeper.

Having DNA is one of three conditions that has to be satisfied for our carbon-based life form to be considered a living organism. DNA contains two long strands which run parallel, but opposite in direction to one another containing many molecules to link them together. There are molecules connecting these strands which contain a person's genes, which constitutes his or her genetic make up. Genes make a person who they are. For this reason some people will have darker skin and why others will have green eyes. The are an infinite number of ways these genes can be organized and every orientation will contain a different person, thus making every individual different.

We have already established that Kirby is not a normal life form. He is comprised of a material we do not understand, but we will assume he has some form of DNA similar to that of a human's. We will assume he is some type of carbon-based life form seeing as he comes onto Earth and breathes, so it is not an unreasonable conclusion. Clearly seeing as Kirby is completely different than our life form, maybe his DNA acts a little bit differently than that of a human.

We can assume that Kirby has regular DNA strands, but has the problem where he has missing connections. Maybe throughout the chain there are protein strands and other connections normally found inside DNA which are missing. So when Kirby is walking around without the abilities of an enemy, he is in an incomplete state. If Kirby is to digest an enemy, maybe the genes of the enemy are then linked to the DNA strands of Kirby, thus why he takes on their powers. So Kirby has a special genetic make up which allows him to single out genes in his enemies which contain their powers and link them to his own genetic makeup.

This could also explain why can loses these abilities as well.

If Kirby is inflicted with too many injuries, he loses the abilities which have become a part of his life. This could very well be because his original DNA is not complete, but stable. So any introductions to his DNA could take the place of the voids, but is not permanent; thus meaning something like inflicting damage could basically knock the newly infused genes out of his DNA. He is also able to dispose of the abilities at any given time. We cannot discredit this seeing as humans are not able to do the acts that Kirby does, we will just have to accept his fantastic ability.

One last ability to examine is the capability of breeding a clone of the enemy Kirby has absorbed. Bacteria can follow through with asexual reproduction, the act of reproduction without a partner. So it is completely possible for Kirby to do the same thing. Kirby is not able to replicate another being exactly like himself, so maybe he has to follow through with sexual reproduction in order to create offspring (do not want to know how!). But when Kirby absorbs an enemy and completes his DNA he is able to create a clone similar to the enemy he absorbed.

When a bacteria reproduces, it makes a clone of all of its information, then sets the clone free which is the process of asexual reproduction. Kirby may follow a similar method, but instead he may store the DNA which is unused from the absorption and combine it all in order to create a clone of the enemy. This would explain why Kirby loses the ability he cloned as he has given it up in order to create an offspring. His DNA is a permanent structure, but the unstable form with enemies DNA is not, so his body would probably prefer to rid of the instability and create an offspring instead.

Further insight about Kirby's composition is provided by Kirby's Epic Yarn, where Kirby is even more minimalistic in design being an outline of himself composed of yarn. Gone are his abilities to suck in enemies and copy their DNA, how can he suck if he has nothing to create pressure (remembering that pressure is caused by molecules bouncing off the wall of a container)? So maybe Kirby's exterior is basically a shell and all of his DNA and abilities are stored within what seems to be his exoskeleton. He can only take advantage of these abilities when he has his suit on in a sense.

Although these concepts do not seem ludicrous to Kirby, they are to the Earth's scientists and life as we know it. We can not conclude that this is impossible, even the life of Kirby himself. Many consider the universe to be without borders, infinitely expanding. Although we have not encountered another life form which acts like what we have on Earth, if the universe is infinite, the odds are in the favour of there eventually being some life form which is similar to Kirby's. It is impossible to draw conclusions one way or another seeing as we are not capable of reaching infinity, so the possibilities are limitless. Since this is such a fun topic, let's apply these concepts to another game we already explored: Pokemon: Soul Silver.

The Pokemon universe is home to hundreds and hundreds of diverse Pokemon to capture. In spite of the plethora of Pokemon we have to choose from, we're really only concerned about an amorphous blob named, Ditto. Ditto is incredibly unique when compared to all other Pokemon in the respect that is does not have any means of attacking, nor a solidified form; it's just a blob. The only move Ditto has at its disposal, and the reason why we are here, is transform.

What Ditto does is use this move on its enemy and assumes almost all of the characteristics of the respective foe. We are talk about all characteristics, form, move set, hatred for being confined to a tiny ball, everything except hit points is replicated. This would explain why Ditto is an amorphous blob, it needs to be able to adapt to conform to the shape of the enemy it is faced with. This would lead us to assume that Ditto shares the same concept of DNA replication and adaptation as Kirby, but at a completely different level. Thus leading us to assume that when Ditto uses transform, almost all of its DNA is replaced with that of the enemies', but there are still some minor parts of Ditto remaining with good reason. After every fight is completed, Ditto will return to blob form, meaning that the same issues of stability that are present with Kirby's case are reflected here as well.

Ditto may not be the most useful Pokemon as the transformation process wastes time and allows it to become a squishy punching bag; but it sure is a neat little Pokemon from an analytical perspective. Even though Kirby and Ditto might have seemingly limitless possibilities, there are cases in a classic RPG series that might not.

We will continue to look at DNA while the idea is fresh in our minds with a wonderful series of games as part of the Dragon Quest side series. The DQ series is what many people claim to be the birthplace of the RPG as we know them today. There were many RPGs before DQ, but it was the first to make them so accessible and enjoyable. They have always been simple games, but their charm have made them one of the most successful franchises ever. Having Koichi Sugiyama, a very prominent Japanese composer, one of the best, at the helm for the soundtracks really bring these games to life. The soundtracks are so incredible that for each main game in the series, a symphonic suite has been recorded from the many live performances. Even though each game's story is simple and straightforward, each is captivating and filled with so much personality. To further praise these titles, each game in the series typically introduces innovative game mechanics, or refines them, like elaborate class systems or capturing monsters. There are many reasons to praise this series, hence why it has captured the hearts of millions and is my favourite game series of all time (probably why I had to find a spot to squeeze it in). Let's press forward with some science!

One of the side series of the DQ franchise is the Dragon Quest Monsters series, each refines the formula used in the previous game, so the explanation applies to all of them. Each game has a main character who has the amazing ability to tame monsters and convince them to aid him/her on his/her journey (more humane then Pokemon) and train and breed them to eradicate other monsters along the way (still not perfectly humane). We will focus our attention on the breeding aspect to see how well it correlates to real life situations. Obviously we will not focus on the actual process as it would be awkward to read and write, we are just worried about the final products.

Much like with animals in real life, the monsters in DQM are contained with a number of different families, each family contains a number of different breeds as well. Let's apply this idea to the most recognizable monster in the DQ franchise, the slime.

The slime is a particular family of monsters. Within this family, there are a number of different breeds, like the heal, king, gran and metal slimes. Each can be thought of much like breeds of dogs, poodles, dobermans, pugs, etc. So it is only logical that when slimes breed with one another, they are receptive to the genetic material of one another, which allows the offspring to inherit characteristics from both parents; characteristics can include special abilities and looks. Some characteristics are more dominant than others, much like in real like. Now that we know this works, what about cross-family breeding?

Have you ever heard of a lippogogamus? A cross between a hippo and a dog. There is a good chance you have never heard of this unless you made it up as well. Ignoring the logistics of how a hippo and dog would procreate, why are one another not receptive to the others' genetic material?

When animals attempt to breed, there are certain conditions that must be satisfied such as similar chromosomal composition (what would dictate having blue eyes or being deaf for example) and compatible love-making material (like a sperm and egg for a human). If these conditions are not met, breeding cannot occur. Sometimes it is an extremely delicate balance, for example, humans are very close in genetic make-up to that of a chimp, but cross-breeding cannot occur. But it can occur in the case of a mule, a cross between donkey and horse. Thus, a lippogogamus is just not feasible, which will apply to DQM.

Even though it is fantastic to cross beast and bird family members with one another to get a bull-bird, the likelihood of this occurring is not very high, probably impossible. Then again, maybe there were some errors at the meeting to determine monster families and all of them are just very diverse forms of slimes, stranger things have happened.

CAMOUFLAGE – MEGAMATOPHORE

Due to the immense amount I love Mega Man and underwhelming amount of MM presented in this book, we had to add another section on this cute little bloke. We will continue to use the original MM game as a reference point, but the explanation holds true for later games in the series as well.

On top of the precise controls, challenging yet fair gameplay, fun Robot Masters (the bosses) and wonderful soundtrack, MM has some really awesome weapons as well. Once MM has defeated a Robot Master, he will assume a particular attack that was frequently utilized during the fight. Now while it would be fun to analyze the weapons, that is not what we are after right now. When MM assumes control of a Robot Master's weapon, his famous blue colours will change to different ones to reflect the weapon change; reds for Fire Man's weapon or greens for Bomb Man's as examples. How or why would MM need to change colours? This is a question proposed by Mike Matei at Cinemassacre, and we are here to heed the call.

First up is the how. Many people are familiar with the chameleon's incredible ability to camouflage; for those less familiar, let's briefly examine it. Depending on the conditions, a chameleon is able to change colours on demand in order to adapt to surroundings. Reasons to adapt can vary greatly, it could be to attract a mate, scare particular predators or hide from others. Regardless of what the reason is, the way to present particular colours is remains the same.

Chameleons have several layers of skin, each dedicated to certain chemical processes to produce desired colours. These layers have chemicals ending with the suffix 'phore', which means to be a carrier. The individual names are complicated and not really of much use to us for the purpose of this explanation. What is important is that each of these 'phore' components work with one another to create a plethora of colours to adapt to every situation. So chameleons basically have the equivalent of a palette of paint under their skin that can be mixed slowly on command.

As indicated previously, MM's technology is incredibly advanced when in comparison to ours, so maybe a way has been found to adapt the biological functions of a chameleon and apply them to robotics; we will call this colour changing layer on MM's armour: the megamatophore layer (best name ever, right?). This megamatophore layer will automatically associate a weapon with a particular colour scheme once it is obtained. This may be a randomly generated, or programmed by MM himself; I don't know, he wouldn't reply to my calls or e-mails. This is all when and good and explains the how, but what about the why?

MM is a very busy little fellow, having to constantly save the world from many dangers, none of which are anything to sneeze at. Due to these dangers, MM probably doesn't have the time to look down at some sort of display on his arm to let him know what weapon he has selected. He may be a robot and it could be automatically linked to him, but he may prefer to keep as few functions running at once in order to be completely aware of his surroundings to ensure victory is achieved. So rather than some sort of display, MM relies on his megamatophore layer to change colours as to let him know what weapon he currently has selected. Thus ensuring he can save the world time and time again, because as the old saying goes: "winners never quit and quitters never win". MM is a winner, and the megamatophore layer is one of many reasons why he is.

PROJECTION – HADOUKEN!

Thanks to one of RetrowareTV's family members Jeremy Pierce (The Gaming Futurist) for suggesting we look at how the Hadouken works. Street Fighter being a seemingly two-dimensional fighting game, developed and published by Capcom. We are not going to bother with any sort of introduction here, look around the internet and you can quickly find out what Street Fighter and the Hadouken are all about. So let's go already!

Let us try and figure out what this Hadouken is composed of. Guess number one is going to involve some electricity. There are very rare instances where balls of electricity can be created. Maybe Ryu has conductors in his hands (painted flesh colour of course) which gather tremendous amounts of static electricity from the air much like rubbing wool socks on a dry carpet. But then again, electricity travels way too quickly and spreads out to the lowest potential spots almost immediately. Those lower potential spots could include surrounding environments and his heart amongst other things. Not only would the ball not be composed quickly enough, it may kill Ryu. Let us Yoga Fire that idea and try a new plan.

The next option is that maybe Ryu pulls powdered copper chloride from his pocket, ignites it and tosses it towards his opponent. For those who are unaware (it really isn't something obvious), copper chloride is used inside of fireworks to create a blue colour after being ignited. Maybe he throws clear balls full of copper chloride to keep the Hadouken shape. This introduces a new problem, the reaction of this powdered metal being ignited occurs in a way too chaotic and quick to be restricted to a slow moving ball soaring through the air. It does not seem physically possible for a Hadouken to be composed by any conventional means. There are very specific conditions which need to be satisfied in order for a Hadouken to be created. There is no way that I can see (perhaps you can, let me know) to create a Hadouken from nothing. I suppose it is time to Shoryuken that idea and try another.

The logical progression is that Street Fighter is something which could be replicated in real life. Stay with me on this one. It would probably be one of the most expensive live entertainment events in history to be composed, but the tournaments are possible. If you have ever seen the hologram technology used to create a Tupac concert, you have your answer as to how this is possible. Each Hadouken, Yoga Fire, Sonic Boom, any energy blast are all images from a highly sophisticated projection system. So not only could Street Fighter be real, it is rigged too. Firstly it would be incredibly difficult to replicate these moves and have them be tracked properly using projectors. Think about Blanka using his electrical moves, those would be very difficult to just project at the drop of a hat, things would need to be thoroughly choreographed. We have more evidence of the tournament being rigged.

If you have played Street Fighter, have you ever noticed how every time you play arcade mode you get to make as many attempts as possible in order to clear it? You get to retry to your heart's desire. Yet your opponent loses once and has to walk away from the situation humiliated in front of the whole world. You get to take your sweet time in order to achieve sweet, tasty victory. You as the player make the most interesting matches seeing as your movements are more random than a programmed AI (unless you are a spammer), so only your winning match will be viewed by the world. The winner is decided prior to even beginning, that is unless you are a quitter (because quitters never win if we recall correctly). Then again, Bison can be quite the pain! How could this be a live event?

For this to happen, there would have to be many projectors, some stationary, others mobile in order to display health, track energy blasts and display character movements. Better yet, maybe the characters would actually be real people and those projectors would create the environments surrounding them. Maybe live Street Fighter is the next logical step to take over professional wrestling (which I don't mind, nothing about pro wrestling appeals to me). Although it would be incredibly expensive, difficult and time consuming to choreograph, Street Fighter may be the future of live entertainment. Jeremy, I think we get a cut of the profits if it does happen.

RESPIRATION – INSPIRATOR

The only reason this section is here is because two of the major characters of Mother 3, Lucas and Claus, have names which are anagrams for one another, and we have a title for this section which follows that quite aptly. For those fortunate enough to be able to speak Japanese or ordered the GBA game and played the fan-translation, you were treated to one of the most unique, fun and unusual gaming experiences of all time. Mother 3 is virtually flawless and any concerns with this game is far outweighed by the unique charm and personality. Mother 3 takes players on an emotional roller coaster ranging from gripping and intense sorrow, to roaring laughter (like when an old man is emulating a monkey dance). There are so many strange sequences and scenes through out Mother 3, what could we possibly be analyzing? How a mailbox can contain 1000 rat corpses? How certain types of dung can be rare and exquisite? No, we are going to analyze the obvious question: how can cross-dressing merman machines give you oxygen?

Surprisingly this is probably one of the most straightforward aspects regarding it. Even more surprising is that the explanation is rather simple as well. First we need to understand how fish that do not go to the surface and get air breathe underwater.

There are some sea dwellers which need to go to the surface to get air such as dolphins and whales. They do not have gills and their bodies run on oxygen with lungs much like humans do, yet there are some fish which stay completely submerged. How do they do this? MAGIC! Section done. Seriously though, as many people know, fish have gills. What these gills are able to is draw available oxygen from the water and breathe like us fleshy humans do with their little lungs. There may only be a tiny amount of oxygen in water (usually in the low single digits percentage-wise), but it is still enough for fish to breathe and live. A similar idea can be applied to our cross-dressing merman machine (I never thought I'd type those words beside one another). Was it mentioned that this game is unusual?

These machines can have some type of filtration system designed in them which only allows oxygen to be drawn from the water and stored until a little boy and his friends come and make out with the machine (possibly forgot to mention this game is unusual). Developing machines which only draw certain particles is quite easy to do. Furthermore, since the machine does not need to breathe like a fish does, the oxygen can be stored indefinitely in our cross-dressing merman machine. So when a little boy decides to be brave and get some oxygen, he can. Other than when they are out of order of course.

There are many factors which can lead to malfunction like there sometimes are in the game. Maybe the filter, probably in the butt following the train of thought of this game, is plugged with some sort of sea creature and can not draw oxygen from the water. Maybe the switch to dispense the oxygen from certain ones has grown deposits of some residue due to travellers not being desperate enough to use each station. I know personally I would take full advantage of the opportunity seeing as it would not come around often in my day to day life and I think you would too; even just to say you did to all of your friends (who will then likely abandon you). Something seems a little fishy about this situation though.

With one of the weirdest games ever created, we have probably one of the most straightforward experiences we have analyzed. Very simple in principle and actually possible in real life. It would certainly be a very terrifying experience travelling in unknown territory not knowing if your next exhale will be your last, but it is possible.

INFINITE LIVES – LIFE IS FLEETING
Much like our extensive coverage of the Metroid series, we are going to cover diverse topics based on suggestions presented by Jordan Fisher from Pwnee Studios. Pwnee Studios developed and published the brilliantly

composed Cloudberry Kingdom just recently, available on just about every platform imaginable! Cloudberry Kingdom is a platforming game taken to all new levels of insanity. Unlike many platforming games, there are several modes of difficulty, ranging from an extremely timid mode to one that even the most masochistic of people would blush at. The fun also never ends due to the random level generator, another very unusual addition to this type of game, or any game for that matter. Enough about the glorious game, onto some physics as we can never seem to back down from suggestions.

First up is the ability to jump ludicrous heights, something we touched on briefly in the beginning of the book (amazing how everything comes together isn't it?). The ability to jump three times our height is rather easy to apply to real life, all we have to do is travel to the moon or go to a facility that replicates the effects of that gravity. This is all well and good for us in real life, provided we have the training and funding to do so, but there is a large flaw with this: people on the moon move very slowly due to the relatively low acceleration due to gravity. Cloudberry Kingdom would be an extremely boring and aggravating game if platforming were slow, hence why it is extremely quick, so how do we solve this problem?

What if someone were born and raised on the moon? If we compare that person to someone who came from earth and jumped on the surface of the moon, the difference in jump heights will be considerably different. The person from earth has had to adapt to a comparatively larger force of gravity, so the gravity on the moon will not be nearly as restrictive. The person from the moon would have not adapted to a greater force than needed, thus the jump heights would be small by comparison. It is similar to a person who can lift 150 kilograms having to lift 10 instead, the training will make the feat much easier than without. We can apply this thinking to two hypothetical planets to solve our problem in Cloudberry Kingdom.

Those doing platforming in Cloudberry Kingdom have all come from planets with extremely large forces of gravity to ones with lesser. This explains why they all fall rather quickly while platforming, yet are able to jump at such ridiculous heights. The next concern asked to be addressed is why a box with a rocket on it, called a rocket box, will refill its fuel when contact is made with the ground.

Combustion can be achieved by burning many types of matter, many gases satisfying the requirements. Oxygen is essential for fires, hydrogen gas is extremely flammable, nitrogen can be used in welding, these are just examples of gases that are combustible. The common thread between them is that they are all found in the air, hydrogen considerably less, but they are all there. So what this rocketbox does, is collect particular components of air through a filtration system, compresses them and then ignites them as some sort of a rocket fuel. This is a continuous process with some type of vacuum system as gases burn extremely quickly, even when compressed, so a reservoir is essential. But then why would there need to be contact with the ground for refueling?

Although this vacuum-compression system is rather effective at collecting and burning appropriate gases, the rate of collection is slightly slower then that of the burning process. So when the rocketbox makes contact with the ground, there are a series of gears that spin due to the friction between the gears and the ground, in order to give the collection system the extra boost it needs to build up the reservoir (similar to the way the Flintstones move their 'car'). Now onto the main attraction.

A major source of pride and a massive selling feature of this game is the masochistic difficulty. Players can actually get rewarded for watching replays with dozens and dozens of deaths, something which seems to be effortless based on the brutal difficulty. Speaking of deaths, this brings up the final suggestion: explain why the player explodes upon death.

We're going to apply the same principle of the self-ingested cyanide pill. There are supposed situations where if people are faced with danger, they will take what is called a cyanide pill which will kill them almost instantly in order to avoid danger. Similar idea is applied to Cloudberry Kingdom with our characters. As soon as they face a threatening situation where death is imminent, the individual platforming will just blow themselves up. Or if they do not have the chance to do so, certain obstacles would have enough force to cause the unstable explosives to be set off anyways. But if the character is dead, why is there another one right there to take over for the probable death run?

We have two possible explanations for this conundrum. The first being that there are an infinite number of parallel universes and each time one character dies, we view one of the parallel universes with a similar character starting the level. This is similar in nature to our discussion about Chrono Trigger. Our second option is that in the Cloudberry Kingdom universe, cloning has been perfected. Each of these clones has the mentality of a lemming and is breed to platform. When you run out of lives, on modes that have them, this means that you have run out of lemming-clones. On the modes that have them, gems are collected as a form of currency to buy more clones. I have no idea how to end this section effortlessly, so, BOOM!

IONS – A SLAVE OBEYS, PHYSICS SHOCKS!

We will be returning to electricity, but using chemistry (an application of physics) to explain why water does not act as a conductor. The idea of water conducting electricity is a very popular misconception which even games have had their fair share of abusing. No game which abuses this idea more than any other, but at least some get the concept correct. So we will examine the idea of why it does not work and what we can do to make it work.

As was stated with Ratchet and his rocket magnet boots, metal is a good conductor of electricity due to its ability to move negative charges through out the respective material. The readiness of a material to allow for the flow of electrons is why certain materials conduct better than others and is what causes a current. Materials such as a block of wood or plastic for example, do not have a ready supply of negative charges to flow through them so they would be considered insulators. Water would fall into the category of being an insulator if it is just plain old water with no contaminants.

Suppose we were to make ourselves a simple circuit with a battery, some wires and a light bulb in a board with two terminals. If we connected the positive terminal of the battery to one of the light bulb's terminals and the negative to the other terminal, we would have a complete circuit which would cause the light to turn on (assuming the light bulb works and the battery has sufficient enough output of course). We can modify this circuit and cut the wire connecting two of the terminals in half and place the halves in a beaker full of water. Doing so, the light bulb no longer turns on. This is due to what was already established about water not being a good conductor. No negative charges are readily available inside of water since the bonds holding the water molecules together are incredibly strong and do not allow for such movement. We can however, modify our water to make it a very good conductor.

If we were to keep the latter circuit described but add salt to the water instead, we would yield completely different results. When salt (sodium chloride) is placed in water the sodium and chloride will 'break' apart into two separate ions, one being positive and the other negative. When looking on the periodic table it is seen that sodium (Na) is on the left side of the table, so it will have a positive charge while conversely chloride (Cl) is on the right with a negative charge. Sodium and all other elements in that column have one more proton than the total number of electrons so they have a plus one charge. Chloride and elements in that column have one more electron than the total number of protons so they have a negative one charge. So as we add more salt there will be more negative charges found throughout the water, more negative charges means electrons can travel to create a current. We will now examine two different cases of water and electricity put together.

First we will observe a case where the water and electricity combination is not having science and the game 'play nice' with one another. Magicka is a fantasy based RPG developed by Arrowhead Game Studios and published by Paradox Interactive. Like many fantasy based games, the player is able to take control of sorcerers which are able to cast spells using a variety of elements to harm their foes. Whether it be the use of water, earth or fire each spell has fantastic effects which can benefit the player in many ways. The use of magic is clearly questionable, but we will allow for the ability to conjure the elements at will, but this exception will not save it from the water and lightning blunder.

A number of combination spells can aid the player in offensive and defensive ways. In terms of the offense, there is the questionable ability to soak an enemy in water to allow for extra shocking power. We know this would not work for two reasons that were introduced before. The first is the more recent one: water does not act as a conductor. The sorcerers in the game do not cast the 'heavily ionized water' spell, so getting their opponents wet would not bring them any sort of advantage to shocking them. The second is brought up with Infamous: electricity will choose the path of least resistance and try to get to a lower energy state. If an opponent were to be standing at certain distances, one of two things could happen. If the opponent is too far away, the lightning spell would simply go into the ground as it is the easier path to take. If the opponent is too close, then it would shock both the user and the opponent, but the water plays no aid in this process. Although it is very fun to soak and shock in the game, it does not obey the almighty physics.

Developed by 2K Boston, Digital Extremes, Feral Interactive and Demiurge Studios and published by 2K Games and Feral Interactive; BioShock is a game which gets the splash and shock mechanics correct. BioShock is a survival, first person shooter RPG which takes place in a failed utopia far in the depths of the ocean. The player must take control of a tremendously confused protagonist which has

to fight strange, mutated enemies called splicers which resemble humans and gigantic, over-protective robot-like beings called big daddies. The protagonist builds a respect-
ably large arsenal of weapons and abilities as the game progresses. Whether it be a crossbow with trip wires or shot-
gun with incendiary rounds, our protagonist has the means
to take on all odds with careful strategy. The observant
reader may have guessed, we will be examining the ability
to shock our enemies.

BioShock developers made an example of pointing
out to the player that if splicers were in a pool of water and
the protagonist was outside of it, shocking the pool is an efficient way to dispose of loathed ones. Unlike the previous
case, the pool of water is actually the path of least resistance. As mentioned before, BioShock takes places in the
depths of the ocean, so we can assume that any of these
pools which the player shocks are filled with salt water.
Seeing as the player is typically on tiles or on some sort of a
non-conductive surface, when the shock ability is used, the
electricity would be more attracted to travelling to the pool
of heavily ionized water as it would be the path of least resistance. If there happened to be any unfortunate souls inside of the pool while the electricity is spreading its energy
throughout the pool, they will become a part of the 'circuit'
as well. Thus if there is enough energy in the initial shock,
this is an incredibly effective tactic to dispose of those
pesky splicers. It may seem cruel to shock splicers to death
but it is the survival of the fittest, hunt, or be hunted. That
is of course if the player chooses to obey the game's suggestions. After all: a man chooses, a slave obeys! (Not a sexist
ending, if you play the game is make more sense)

FINAL THOUGHTS

Thus ends our relatively short journey. I would like to thank you, the reader, for seeing the book through this far. I appreciate your support and the support of everyone who was involved in this. There have been so many suggestions that I am extremely grateful for, thank you believing in this work. I hope it has been a source of enjoyment for you as it was a pleasure to write.

I hope this has been an insightful and interesting demonstration of how physics and mathematics can be found in something as seemingly simple, but clearly complex as playing a video game. I hope this has inspired you to learn more, do your own analyzing, create your own content in something unrelated, anything. Follow your passion, let nothing stop you from doing what you love.

Thank you very much for the love. Now if you'll excuse me, I have to contort onto a floating chocolate and drift off into a dream.

REFERENCES AND ACKNOWLEDGMENTS

Special thanks to:

Dexter Morrill for designing/drawing the cover: http://dextermorrill.wordpress.com/.

The Commodore for helping edit and write the section of Chrono Trigger: www.clanofthegraywolf.com.

MatPat of Game Theory who inspired me and allowed me to offer extensions of his work regarding Samus' morph ball. Also for inspiring me to write the Game Theory user submission for Portal 2 (never got used to my understanding): https://www.youtube.com/user/MatthewPatrick13.

James Kakalios for inspiring me with his book and lecture I attended, he was a driving force for me to complete this work.

Arin Hanson (Egoraptor) for helping me see how much a game can teach a person without forcing tutorials upon them (never met him or talked to him about this).

Roo from the Clan of the Gray Wolf, who inspired me with his well-received scientific video game explanations: www.clanofthegraywolf.com / www.retrowaretv.com

Those who continue to support me at RetrowareTV, whether it be hosting (thanks everyone who is involved with that!) or reading my work: www.retrowaretv.com.

LaserFrog of *Run, Play, Think!* for allowing me to emulate him in the section with Batman: http://retrowaretv.com/category/shows/runplaythink/.

Those who inspired, supported and/or made suggestions for me to cover:

Michael Anthony of *Still Loading* – Nights Into Dreams, Super Paper Mario.

Clint of *Lazy Game Reviews* (http://www.lazygamereviews.com/) – Conker's Bad Fur Day.

Brent Blauser of *Gaming Legends* – Mega Man X Street Fighter, Mega Man X and Tomba!.
Marc Carr of *Indie Games Searchlight*
(https://www.facebook.com/pages/Indie-Games-Searchlight/119133874766675) – Ascension.
Pat Contri of *Pat the NES Punk* (http://www.thepunkeffect.com)
– Duck Tales.
Ben Hall of *Video Game Take-Out*
(https://www.facebook.com/VideoGameTakeOut) – Batman,
Umihara Kawase.
LaserFrog of *Run, Play, Think!* – Batman.
Jeremy Pierce of *The Gaming Futurist* – Street Fighter.
All of them (except Marc, who helps there) can be found at:
http://www.retrowaretv.com.
My family, friends and fans (I wish I knew a better word, that seems awkward) for helping and supporting me with this project. Their patience is much appreciated. I will stop asking you for suggestions now (for a little while anyways).
Victoria for being my constant source of inspiration and encouragement.

TEXT AND WEBSITE

Beiser, A. (2003). *Concepts of modern physics*. New York, NY: McGraw-Hill Education.

Fowles, G., & Cassiday, G. (2005). *Analytical mechanics*. (7 ed.). Belmont, CA: Thomson Brooks/Cole.

Griffiths, D. J. (2005). *Introduction to quantum mechanics*. (2nd ed., p. 430). Upper Saddle River, NJ: Pearson Prentice Hall.

Kakalios, J. (2009). *The physics of superheroes*. London, England: Penguin Books.

Scherrer, J. (2010). *U.S. rope and cable*. Retrieved from http://www.us-rope-cable.com/

U.S. Army (2010). *Black hawk fact file for united states army*. Retrieved from http://www.army.mil/factfiles/equipment/aircraft/blackhawk.html

VIDEO GAMES (IN ORDER OF APPEARANCE)

I am trying my best to give credit to the respective owners, if there are concerns please contact me and I will adjust the credits accordingly.

Cover

Drawn by Dexter Morrill, the cover features components from the following games and are credited in this section: the background and blocks are from the Mushroom Kingdomin Super Mario Bros., floating chocolate is from Conker's Bad Fur Day, boots are from Mega Man X, pants are from Banjo Kazooie, pokeball is from Pokemon Soul Silver, shirt and cape are from Fire Emblem: Radiant Dawn, glove is from Batman, arm cannon and wave beam are from Metroid, headband is from Metal Gear Solid and the face is my charming face as a cartoon.

Forces and Motion

Super Mario Bros. was published and developed by Nintendo in 1985

Contra was published and developed by Konami from in 1987
Mega Man was developed and published by Capcom in 1987
Batman: The Video Game was developed and published by Sunsoft in 1990
Mega Man X was developed and published by Capcom in 1993
The Elder Scrolls: Skyrim was developed by Bethesda Game Studios and published by Bethesda Softworks in 2011
Ninja Gaiden Sigma 2 was developed by Team Ninja and published by Tecmo Koei in 2009
Mortal Kombat was developed by NetherRealm Studios and published by Warner Bros. Interactive Entertainment in 2011
Flight
Fire Emblem: Radiant Dawn was developed by Intelligent Systems and published by Nintendo in 2007
Banjo-Kazooie was developed by Rare and published by Nintendo in 1998
Gravity Rush was developed by Japan Studio and published by Sony Computer Entertainmentin 2012
Sound
Dead Space was developed by Visceral Studios and published by Electronic Arts in 2008
Projectile Motion
Call of Duty: Modern Warfare 2 was developed by Infinity Ward and published by Activision in 2009
Tension
Call of Duty: Black Ops was developed Treyach and published by Activision in 2010
The Legend of Zelda: Link's Awakening was developed by Nintendo EAD and published by Nintendo in 1993
Harmonic Motion
Metroid I-III were developed and published by Nintendo

from 1986 – 1994

Bionic Commando was developed and published by Capcom in 1988

Friction

Assassin's Creed II was developed by Ubisoft Montreal and published by Ubisoft in 2009-2010

Shadow of the Colossus was developed by Team Ico and published by Sony Computer Entertainment in 2005 & 2011

Centripetal Motion

Crash Team racing was developed by Naughty Dog and published by Sony Computer Entertainment America in 1999

Mechanical Advantage

Crysis 2 was developed by Crytek and published by Electronic Arts in 2011

Elasticity

Mega Man X Street Fighter was developed and published by Capcom in 2012 and can be found as a free download online on the Capcom Unity website

Duck Tales was developed and published by Capcom in 1989

Umihara Kawase was developed by TNN and published by someone (I seriously can't find out who, please let me know and I will fix this) in 1994

Ideal Gas Law

Kirby Super Star was developed by Hal Laboratory and published by Nintendo in 1996

Super Smash Bros. Brawl was developed by Ad Hoc Development Team and published by Nintendo in 2008

Donkey Kong Country 3: Dixie Kong's Double Trouble was developed by Rare and published by Nintendo in 1996

Dokuro was developed by Game Arts and published by GungHo Online Entertainment in 2012

The Legend of Zelda: Oracle of Seasons was developed by

Capcom and Nintendo and published by Nintendo in 2001
Nights Into Dreams was developed by Sonic Team and published by Sega in 1996
Tomba! was developed by Whoopee Camp and published by Sony Computer Entertainment America in 1998

Activation Energy

Atomic Bomberman was developed and published by Interplay in 1997

Electricity

Infamous I & II were developed by Sucker Punch Productions and published by Sony Computer Entertainment in 2009 & 2011

Magnetism

Ratchet and Clank Future Trilogy was developed by Insomniac Studios and published by Sony Computer Entertainment in 2007 – 2009

Conker's Bad Fur Day was developed and published by Rare in 2001

Magnetic Fields

Silpheed: The Lost Planet was developed by Game Arts and Treasure Co. Ltd. And published by Working Designs in 2001

Electromagnetic Radiation

Metal Gear Solid I-IV were developed by Kojima Productions and Konami and published by Konami from 1998 – 2011

Ascension was developed and published by Magnesium Ninja and can be found for free, here: http://www.magnesiumninja.com/

Quantum Tunnelling

Pokemon Soul Silver and Heart Gold were developed by Game Freak and published by Nintendo and The Pokemon Company in 2010

Time Travel
Chrono Trigger was developed and published by Square in 1995
Dimensions
3D Dot Game Heroes was developed by Silicon Studio and published by Atlus in 2009
Super Paper Mario was developed by Intelligent Systems and published by Nintendo in 2007
Teleportation
Portal 2 was developed by Valve Corporation and published by Electronic Arts and Valve Corporation in 2011
Eat Lead: The Return of Matt Hazard was developed by Vicious Cycle Software and published by D3 publisher in 2009
Wavefunction Paradox
Final Fantasy VII was developed by Square and published by Sony Computer Entertainment America in 1997
Contortion and More
Metroid I-III were developed and published by Nintendo from 1986 – 1994
DNA
Kirby Super Star was developed by Hal Laboratory and published by Nintendo in 1996
Kirby's Epic Yarn was developed by Feel-Good and Hal Laboratory and published by Nintendo in 2010
Pokemon Soul Silver was developed by Game Freak and published by Nintendo and The Pokemon Company in 2010
Dragon Quest Monsters series was developed by TOSE and published by Enix/Square-Enix from 1999-2013
Projection
Street Fighter II was developed and released by Capcom in 1992
Respiration
Mother 3 was developed by Nintendo, Brownie Brown and HAL Labratory and published by Nintendo in 2006

Infinite Lives
Cloudberry Kingdom was developed and published by Pwnee studios in 2013
Ions
Magicka was developed by Arrowhead Game Studios and published by Paradox Interactive in 2011
BioShock was developed by 2K Boston, Digital Extremes, Feral Interactive and Demiurge Studios and published by 2K Games and Feral Interactive in 2007